T0190033

Map Framework

Mark McKenney · Markus Schneider

Map Framework

A Formal Model of Maps as a Fundamental Data Type in Information Systems

Mark McKenney
Southern Illinois University
Edwardsville, IL
USA

Markus Schneider
University of Florida
Gainesville, FL
USA

ISBN 978-3-319-83580-8 ISBN 978-3-319-46766-5 (eBook)
DOI 10.1007/978-3-319-46766-5

Softcover reprint of the hardcover 1st edition 2016

Printed on acid-free paper

This Springer imprint is published by Springer Nature
The registered company is Springer International Publishing AG
The registered company address is: Gewerbestrasse 11, 6330 Cham, Switzerland

To my family, Elizabeth, Samantha, and Evelyn, for their support and understanding through the hours of research and writing that went into it. And to my parents, Scott and Joan, for their never-ending support and encouragement.

—Mark McKenney

To my family, Annette, Florian, and Tim, as well as my parents Hans and Christel Schneider, for their never-ending love, support, and encouragement.

—Markus Schneider

Contents

Acronyms

9IM	9-Intersection Model
DBMS	Database Management Systems
GIS	Geographical Information System
LPNP	Labeled Planar Nodeless Pseudograph
PNP	Planar Nodeless Pseudograph
SSH	Spatial Semantic Hierarchy
SSPG	Structural Spatial Partition Graph

Chapter 1
Concepts of Maps

The term *map* carries a variety of definitions under multiple contexts; however, the notion of a map is familiar to people despite the many meanings of the term. Maps are fundamentally tied to notions of location and space, but a map as a tool to communicate the location of objects is but one use of the concept. More advanced uses of maps include:

1. The integration of different types of information into a single, expressive source.
2. The communication of spatial and topological relationships among items.
3. The reduction or exaggeration of spatial attributes to emphasize aspects of information.

For example, a map may integrate topography information with road networks. Although the two types of information are quite different, their integration into a single source provides complex information, such as the steepness of roads, in an expressive form. Basic spatial concepts such as proximity of objects and length of routes are easily depicted in maps. Finally, color and scale are often used to emphasize informational aspects. An example is a map of countries where brighter colors indicate higher populations. Despite the various conceptions and implementations of maps, the common thread is that maps tend to be tools of information expression, and they are used effectively in this role precisely because of the variety of types of maps. Furthermore, maps tend to encode information in a fundamentally visual manner. Thus, complex information is reduced to a visual form that is readily understood by users of maps.

In fact, it is the visual nature of maps that often defines their use. Users of maps typically *look* at maps in order to derive information from them. This notion of maps as visual tools has carried over into the computational realm. Geographic information systems (GISs) are software systems that store and provide analysis capabilities for *spatial information*. GISs have the capability to produce many types of maps. However, GISs produce maps as visual entities, especially when using data sources that express spatial data in a vector format as opposed to a raster format.

M. McKenney and M. Schneider, *Map Framework*,
DOI 10.1007/978-3-319-46766-5_1

In other words, the maps are the product of spatial analysis, rather than the data upon which computational analysis is performed. Instead, spatial data is stored and manipulated in more primitive elements, and combined to visually create a map as an output product. Again, the concept of maps as visual tools is maintained.

In this book, we explore a new conception of maps that breaks away from the role of maps as visual tools. We propose a concept of a map as a fundamental data type that can be stored in information systems, manipulated through operations, and combined with other maps. The term *data type*, in general, indicates a set of possible values that a particular data item can store or represent, and (equally important) restrictions on what values a particular data item cannot store or represent. Furthermore, a data type indicates what operations may be performed on instances of the type, and what the semantics of those operations are. A data type forms an *algebra* since it specifies both data elements and a set of operations for those elements. We say that a data type is a *fundamental data type* or a *first class citizen* in a system if it is natively supported by the system. For example, a programming language may define a fundamental data type of unsigned integers as all positive integers represented by, at most, 32 bits in binary format. The language could then provide operations to add, subtract, multiply, etc., unsigned integers. A user could then create a new data type of rational numbers consisting of an unsigned integer as a numerator and an unsigned integer as a denominator; however, without a rigorous type definition of what is allowed and what is not allowed, and a specification of a complete set of operations and semantics over rational numbers, the new data type cannot be integrated as a fundamental data type into the language (for instance, is the denominator allowed to be 0?). Thus, the user created type is not natively available to all users of the language, and there is no standardized data type of rational numbers across the language.

Current implementations of maps do not fit the notion of a fundamental data type, and in many cases, are treated more as visual artifacts rather than data types; such conceptions of maps limit the functionality of maps in computational systems. This is not to say that a user created data type cannot be considered a fundamental data type, but one must rigorously define the data type and integrate it into a system such that it may be used seamlessly just as any other fundamental data type. Therefore, we propose a fundamental data type of maps, along with a collection of operations that manipulate maps and produce new maps. Thus, we propose a *Map Framework* to directly store and manipulate maps as first class citizens in information systems. Because maps have the ability to inherently represent complex information, the Map Framework provides powerful data analysis opportunities in a concise, algebraic formulation. Furthermore, we consider maps in the context of vector representations of spatial data, rather than raster definitions. Again, the multiple meanings of the term "map" come into play when discussing spatial data since data in raster format are often referred to as "images" or "maps" and are again differentiated from the more visual conceptions of maps. In the remainder of this chapter, we describe concepts of maps in a variety of fields; the following chapters build the Map Framework that provides the data types and operations necessary to implement the conceptual components of maps used in those fields.

1.1 An Idea of Maps

Because maps are such ubiquitous constructs, we take the view that a map should be able to be represented as a fundamental data object. For example, we should be able to define a fundamental data type of maps such that maps can be stored as *first class citizens* in databases. The implication of this conception of maps is that one can execute queries over maps in order to analyze them, or create new maps from existing maps by combining existing maps through a composition of predefined operations.

As an example of the idea of maps as a fundamental data type, consider a database containing maps of forest coverage. Each map contains multiple areas that define the predominant type of plant growth, e.g., tree cover, grass cover, shrub cover. Each map may represent the same general area, but at a different time. Under the conception of maps as visual tools, a user can compare changes in plant growth by either visually examining various maps, or by performing operations on the underlying data used to create the map. However, because the underlying data is treated as separate instances of fundamental data types, such as integers, performing complex operations over the maps may be very difficult or beyond the capabilities of a system. If the maps were first class citizens, or fundamental data objects in themselves, they could be combined using operations built into the system through, for example, a query construct. Furthermore, such queries could operate on large numbers of maps that may overwhelm visual analysis. In this book, we are concerned primarily with vector representations of spatial data, as opposed to raster representations.

1.2 Towards Maps as a Fundamental Data Type

Current GIS systems are designed around the concept of maps as visualization tools rather than maps as data objects. One reason for this is that spatial systems represent spatial data, particularly vector spatial data that we are concerned with, as more primitive constructs; in particular, a spatial datum in vector format is typically stored as a point, a line, or a region (Chap. 2). A map must combine each of these primitive objects in order to represent a more complex, yet still fundamental, data type. In this book, we present a formal type definition for a data type of maps that can be used as a fundamental data type in spatial systems. We then develop fundamental operations over maps that may be used to construct complex work flows for map manipulation and querying. We show how those operations can be combined to implement map operations described in the literature. Finally, we provide topological relationships between maps that can be used for topological queries.

Over the course of this book, we build the Map Framework in steps. The first step is to define the *abstract model* of maps. The abstract model is a mathematical formalization of a map data type along with mathematical definitions of operations over maps. At the abstract level, we are concerned with creating a precise data type for which we can prove type closure under operations; in other words, we show that map operations will produce maps as output such that operations can be composed to define complex data analysis tasks. At the abstract level, we do not consider implementation aspects, thus, we will take advantage of concepts such as infinite point sets that cannot be directly implemented in computer systems, and we will not consider time or space complexity for operations. The focus at the abstract level is to create a mathematical foundation for the Map Framework.

Once the abstract specification is complete, we move on to the *discrete model* of maps. At the discrete level, we translate the abstract data type for maps into discrete constructs that can be implemented in computer systems; however, we do not consider the implementation of the model in a particular system. In other words, the discrete model is independent of implementation particulars. For example, the discrete model of maps does not impose a particular numerical data type to represent coordinates, rather, this is left to the implementation model.

Finally, the *implementation model* of maps provides mechanisms to implement the discrete model of maps in a particular system or setting. In this book, we provide an implementation model geared towards databases. Thus, we discuss mechanisms to store maps and their attribute data in databases and show how to implement map operations specified in the abstract model in the database setting. One powerful aspect of this sequence of defining maps is that a common abstract data type that provides the exact specifications for the map type and the expected behavior of operations can be implemented in many settings and in many ways, yet all implementations will have the same type semantics.

1.3 Fundamental Components of Maps

Because the term "map" has such varied definitions and connotations, we begin by specifying the fundamental characteristics of maps and explore maps in various contexts with regard to those characteristics. At its most basic level, a map contains three fundamental constructs:

- Representation of space.
- Representation of topology.
- Representation of thematic values.

As we will see, these three constructs are available in nearly all applications of maps.

Spatial representation in maps occurs primarily through drawing of objects. For example, points are frequently used to represent landmarks, lines represent roads and rivers, and lines are joined to enclose areas representing regions. These three types of geometries are sufficient to represent a vast array of items on maps. However, maps do not need to represent solely geographic reality. For example, maps may tend to exaggerate reality in order to make a point. For example, a map of the United States that is emphasizing information about the state of Illinois may show that state as much larger than the other states in order to emphasize some point. Therefore, it is important to remember that although maps may be geographically grounded in many instances, the goal of a map is not necessarily to accurately reproduce geographic reality. A fundamental map data type must be able to represent geometry; thus, there should be no restrictions that the geometry be specifically tied to geography, although it can be.

Topological representation is inherent in many forms of maps and lends itself naturally to visual representation. Topology also forms a fundamental class of queries for analysis of objects within maps, and over maps themselves. Topology describes qualitative relationships among objects while ignoring quantitative relationships. For example, the US states of Illinois and Florida are topologically disjoint; they are no more or less disjoint than Illinois and Texas, they are simply disjoint. Thus, the distance between spatial objects is not of concern, as long as that distance is non-zero. From a user's perspective, topology is fundamentally tied to navigation. In some sense, navigation is a sequence of topological relationships among landmarks. For example, to travel from one city to another, a traveller encounters a sequence of topologically connected constructs: a city, then a highway, then a city, etc. Distance comes into play to calculate duration of travel, but the route itself is a series of landmarks sequenced by topological relationships. This notion of topology is key to many types of maps.

Thematic values are used in maps to associate data with the spatial and topological aspects of the maps. Thematic values can be any type of data associated with an aspect of a map. Familiar types of thematic values are names of countries or roads, populations of cities, elevation of a topographic contour, etc.

In the remainder of this section, we examine concepts of maps from various disciplines. This examination shows that the three fundamental constructs of space, topology, and thematic data are sufficient to describe the map concepts in these varied fields. Our proposal of a fundamental data type of maps will implement these three constructs and form a construct upon which maps in the fields discussed below can be implemented, stored in computer systems, and analyzed using a general algebra that is not tied to any one field.

1.3.1 Types of Maps

In the literature, three key terms arise to classify particular kinds of maps:

- graphical or geometric maps
- cognitive maps
- sketch maps

Graphical maps describe the classical connotation of a visual map that has been in use for centuries, such as a map on paper (although graphical maps may be displayed by computing devices as well). Traditional road maps and maps in atlases are common implementations of graphical maps. The goal of graphical maps is often to relay information. For example, road maps are meant to relay relative distance, direction, and routes between destinations; heat maps relay information about the relative density of some attribute, for example a map where areas of high crime have more intense color than areas of lower crime. In our discussion, we use the term *geometric maps* to refer to graphical maps since graphical maps are primarily composed of various geometries arranged for visual consumption; however, the term "geometric maps" is more broad since it does not imply visualization as the primary goal.

The term "cognitive maps" has many meanings, just as the term "map". We refer to cognitive maps using the metaphor of a *map in the head* [7, 18–22, 24, 25, 37]. In other words, a cognitive map is the knowledge of a large-scale space as conceptualized by a human. Cognitive maps differ from graphical maps in the sense that they are typically constructed incrementally from observations over time. Cognitive maps, in this sense, have clear applications in robotics where a robot must sense its environment in some manner in order to construct a map that allows it to navigate its surroundings. An interesting aspect of cognitive maps is that they are not necessarily spatially grounded, instead they can be purely topological representations. For example, a cognitive map may simply include landmarks in relation to each other, the landmarks are not necessarily grounded in spatial coordinates or even include concepts of distance.

Sketch maps refer to maps sketched by humans. For example, a person who must provide directions to friend might sketch a map on a piece of paper (hence the name). Sketch maps incorporate aspects of both cognitive and graphical maps in that they are visual, but are frequently spatially distorted in order to emphasize topological relationships. For instance, a sketch map of a geography containing a road that runs for ten miles, then makes a turn at a water tower, then proceeds five miles to a destination may show two road segments of equal length, but emphasize the turn at the water tower by drawing a large picture of the water tower [27].

Interestingly, topology and attribute values are key elements of each of these maps, but spatial information is not a requirement of cognitive maps. This may be somewhat counterintuitive given popular connotations of maps, but emphasizes the point that the benefit of arranging data into map form arises from the topological information that is either explicitly or implicitly represented in the structure. Various disciplines use some or all of these types of maps, but each has a focus on topology.

The remainder of this section discusses the use of maps in various contexts. The purpose of exploring maps in various contexts is to show that maps under all these contexts rely on the fundamental properties of representation of space, topology, and thematic values. In the following chapters, we will develop a Map Framework that incorporates these properties into a fundamental data type; therefore, the Map Framework represents a data type that can be used to implement maps in the various contexts explored in this section.

1.3.2 Maps in Robotics

Maps form one possible mechanism for robot navigation. Although there is vast work on implementing map-based robot navigation, we focus on the conceptual work.

Robotics provide a unique way to study maps and their use because a robot can truly begin exploration of its surroundings with zero initial knowledge, as opposed to humans who have a history of locating themselves in regards to their surroundings. Therefore, robotics provides a unique way to test the question of what type of map is truly needed for navigation. For example, is a cognitive map containing purely topological information sufficient for locating oneself in an environment and then navigating to a destination, or does spatial information need to be present as well? Furthermore, is it more effective to build a cognitive map starting with topological information and ending with spatial information, or is it better to begin with spatial information of ones surroundings and then derive or build the topological information?

References [24, 25] proposes a *Spatial Semantic Hierarchy* (SSH) that builds a cognitive map from sensory input and ends with the construction of *metric maps*. Under our terminology, a metric map is a map in which distance is defined, and falls under the category of geometric maps (even if a graphical representation is not visualized). Map construction under the SSH model proceeds with the following sequence of *control levels*:

$$\textit{sensory} \rightarrow \textit{control} \rightarrow \textit{causal} \rightarrow \textit{topological} \rightarrow \textit{metrical} \tag{1.1}$$

The *sensory* control level is where input to the robot is taken through sensors. The *control* control level binds the robot and its environment to create a dynamic system. A local geometric description of the robots surroundings and goal, such as hill climbing or line following, can be coupled to enable a control system to address some goal. The *causal* control level consists of models that describe the causal relations among "views" of sensor data and actions. For example, measuring turn magnitude to complete a turn action is dependent on sensors that indicate how far the robot has turned. The *topological* level includes descriptions of paths, places, regions, and connectivity. According to [24], topological maps are effective for planning purposes. Finally, the *metrical* control level introduces a global geometric map of the environment.

The SSH model clearly is attempting to create an abstract hierarchy that includes low level sensory needs of robotic agents; yet it provides interesting insight into one mechanism of using maps; specifically, the SSH model indicates that the topological level is the most important for robot planning, and that the metrical level is not always essential. This is a notion that holds true with human cognitive maps as well. Again, the key point is that the topology is an essential component to maps

Reference [17] again highlights the importance topology in maps, but does so by building global topological maps form local geometric maps. The central idea is to build a global topological network of local geometric spaces. The observation made is that a robotic agent can readily create a geometric map of its local surroundings for navigation of the immediate area, but must be able to topologically integrate that local geometric map into a network of other geometric maps. This avoids the problem of having to merge individual small geometric maps into a single large geometric map. In this approach, geometric maps play a more necessary role, but it is the topology that provides the expansion of information into larger spaces.

Other approaches to conceptual mapping in robotic agents exist, but the emphasis on topology as the core element with geometry forming a supporting element is typical [8, 22, 23, 38]. What is not explicitly stated, but implied, is that thematic attributes are present for higher level navigation. An absence of thematic information is sufficient for obstacle avoidance, but higher level planning requires more data about one's surroundings.

1.3.3 Cognitive Maps

Robotic mapping draws heavily from concepts in human cognitive mapping. Although research into human cognitive mapping quickly leads out of the scope of this work, early work on human cognitive mapping and the development of cognitive maps, the work that robotic mapping derives much from, is applicable. Early work in the area [33] investigated whether cognitive maps developed by rats for finding food were "strip maps", meaning they contained no context outside of the turns that must be made to achieve a goal (food), or whether the maps were "broad", containing larger context. The conclusion was that cognitive maps in rats contained properties of "strip maps", meaning that the strip maps where cognitively present, but there were signs of the maps being embedded in some limited "broadness". One can see the implementation of this concept in the robotic realm where larger maps are constructed from local perceived geometries or from local topological awareness.

Later work built upon the early ideas of generalizing local cognitive maps into broader maps. Reference [1] looked at the ability of 3, 4, and 5 year old children to generalize the position of a goal (a trinket in this case) after having been placed in a different orientation to the one in which the location of the goal was originally made known to them. Different experiments looked at performance if land marks were provided and if models of the space were provided. Again, the central question is whether the subjects could coordinate their local conception of space with large-scale conceptions of space despite changes in orientation. The introduction of landmarks were considered as providing topological information to the subjects, and allowed even the youngest children to compensate for the change in orientation. Again, importance of topology in maps cannot be understated.

An interesting thought when considering cognitive maps is: "what does a materialized cognitive map look like?" One way to materialize a cognitive map is to draw a *sketch map* [4, 5, 27, 28]. Indeed, sketch maps may be considered a manifestation of a cognitive map. Sketch maps provide interesting insights into cognitive maps. For example, sketch maps are often geometrically simplified, can be inaccurate in terms of distance, but tend to be accurate in terms of topology. This again underscores the importance of topology in maps.

1.3.4 Geometric Maps

At this point, we move from the more abstract concepts of maps to more concrete map representations. We examine geometric map representation from a computational perspective, since our aim is to develop a model of maps as a data type. Computational notions of maps emerge from geometric conceptions of maps. Geometrically (that is, ignoring attribute values of map elements for the time being), maps are based on a *partitioning* of some embedding space. We assume the embedding space will be the topological space $\mathbb{R}^2$. A topological space is a set X ($X = \mathbb{R}^2$ in our case) with a collection of open subsets of X:

$$U \subseteq 2^X \tag{1.2}$$

such that X and U satisfy the following axioms:

1. $\emptyset$ and X are in U.
2. The union of any members of U belongs to U.
3. The intersection of any finite number of open sets in U belongs to U.

Because set difference can be expressed in terms of intersection and union, it follows that the type of open sets is closed under intersection, union, and difference.

Therefore, the partitioning of a topological space $\mathbb{R}^2$ requires the definition of a *topology* U upon the space. Furthermore, it is often useful to define a *basis* β for a topology such that every element of U is equivalent to a union of elements in β. A *partition topology* is defined as a topology such that elements in the basis for that topology are disjoint.

Thus, a geometric map is a topological space together with a partition topology. In the topological space $\mathbb{R}^2$, this implies that the embedding space is subdivided into non-overlapping *regions*. In practice, such regions are often represented in computational settings by storing collections of straight line segments that represent the boundaries of the regions.

Representations of geometric maps in the computational realm have been used for a long time. The winged edge structure [2, 3] provides a graph based data structure to represent a partition and its topological structure in a single data structure. The data structure provides an algebra of operations that allow one to *walk* the edges and nodes in the structure. For example, two regions that share a boundary segment share an edge, and that edge contains information to identify those regions. Points where three or more regions meet create nodes that allow access to the edges that emanate from the node, as well as the faces that border the node. Thus, one can imagine a bug walking the graph and using the various operations to identify the next and previous node, edge, etc. Furthermore, because each segment identifies the faces it bounds, the winged edge structure inherently represents the dual graph of a partition.

Winged edge structures are useful in that they successfully merge the idea of an algebra with a data structure, the algebra is focused on manipulating and traversing a single instance of the structure, and not on analyzing and manipulating groups of maps. Thus, the pure structure is not appropriate for our goals, but the many of the ideas pioneered by the winged edge structure influence our proposal.

The doubly connected edge list [30] shares many of the concepts of the winged edge structure, including the notion of traveling a graph representation through operations and inherently representing both a geometry and its dual graph. However, the doubly connected edge list is developed to compute the intersection of convex polyhedra in 3D space. We focus on partitions in 2D space, so the pure data structure itself is not applicable. Again, many of the concepts used in the doubly connected edge list influence our work.

Another important data structure in the realm of geometric maps is the quad edge data structure [11]. Again, the quad edge structure is a data structure that represents a partition, inherently represents the dual graph of the partition, and provides an algebra to walk along the graph and dual graph and interact with the regions and segments of the data structure. Although the quad edge structure is designed with voronoi diagrams in mind, it is not necessarily tied to them. The quad edge structure is not applicable to our situation due to the reasons listed for the other structures in this section, but it also influences our work.

1.3.5 Thematic Maps

Thematic maps are centered around the association of data with elements in a map. For example, one may associate a name with each region in a geometric map in order to create a thematic map. In general, the geometric structures in a thematic map are defined based on their thematic information in the sense that an area with an identical thematic value forms a *region* in the map.

Although a thematic map may be implemented in a variety of forms, one of the most widely used forms in term of computation is the *raster* map or raster image. More generally, a raster map is a *tessellation* approach to representing maps. Tessellation approaches impose a tessellation scheme on the underlying space and assign a value to each cell; this value is termed an attribute value. Attribute values may be numerical, text, or any other value. Adjacent cells with identical attributes form *fields* or *regions*. Raster maps are the subject of much study in the literature, resulting in algebras of raster maps in multiple dimensions that provide much functionality [10, 26, 29, 32, 34]. Raster maps also bridge the ides of maps as data elements and maps as visualization tools: a raster map is inherently tied to the concept of an image and raster visualization is straightforward. The main drawback to raster maps is that they introduce difficulties when incorporating data in non-raster formats. Much data is expressed in terms of spatial data objects in vector formats, in which the boundaries of objects are represented (usually as linear structures) as opposed to the raster approach of effectively representing the interior of an object as a collection of adjacent, identically labelled cells. Vector representations are appropriate in many instances. Although vector data can be converted to raster, it can result in (i) very large data due the required storage of many cells to fill the interior of a region, (ii) a loss of precision due to the fixed cell size in a raster image, and (iii) and introduction of a fixed resolution of the data. Such drawbacks are clear if one considers a rectangle. In a vector format, the line segments defining a rectangle may simply be represented by four end points (or two, for that matter), whereas a raster image must represent pixels for the entire interior of the rectangle. A loss of precision may occur because the boundary must effectively be "rasterized", which may lead to pixelation. This is especially problematic if a region boundary is curved, non-rectilinear, of if two regions share a boundary: their shared boundary must be pixelated so as to not introduce overlapping portions of the regions. Finally, because a region must be pixelated, it is pixelated at a particular resolution; this complicates matters when combining data from multiple sources in multiple resolutions. Although these problems can be remediated by various means, they remain an issue that must be addressed in practical applications.

Tomlin [34, 35] developed a map algebra for raster maps. Tomlin's map algebra is significant in that it formalizes a complete set of types and operations over raster maps in which type closure guarantees are provided; thus, the output of one operation can be used as input for another. In general, operations are considered to be *local, focal, global,* or *zonal*. The class of operations indicates the scope of the operation: local operations affect individual cells, focal operations affect cells and their neighbors, global operations affect an entire raster image, and zonal operations affect regions (i.e., neighboring cells that share the same value). Tomlin's algebra forms the foundation of raster processing systems, but is fundamentally tied to the raster context. Thus, the need for a general map algebra over vector maps exists.

Vector maps, such as geometric maps, may be used for thematic maps. In this case, thematic data is associated with the geometric structures in the map. This approach informs the *collection* approach to map modeling. The collection approach takes the view that a map is simply a collection of more basic spatial entities that may satisfy certain topological constraints. This is the approach taken in [9, 12, 13, 16, 31, 36]. In general, collection types do not implicitly support thematic maps, but model maps as purely geometric structures. Some models provide mechanisms to associate thematic data with the individual components of the map, but the thematic data is not part of the map definition. Furthermore, no type closure is provided. Additionally, no method of enforcing topological constraints over map contents is provided in the model itself. Geometric map structures, as discussed earlier, are one approach to a collection map, however, collection maps generally store the geometric primitives that define spatial structures in the map (i.e., points, lines, and regions in the map) as individual geometric objects. By storing geometric primitives as individual objects, the map itself is not a fundamental data type in an information system, rather the map is a visualization construct or a set of constraints imposed on a collection of the underlying data objects.

The Topologically Integrated Geographic Information System (TIGRIS) [6, 14, 15] is unique in that is an early system that incorporated maps into its spatial data storage model. The TIGRIS system is designed around the collection model of maps, such that individual geometric primitives form a map which contains no overlapping regions. For example, if a region representing Florida and a region representing a hurricane that partially overlaps Florida are stored, then three regions are actually stored by TIGRIS, the region representing the intersection of Florida and the hurricane, the region consisting of the part of Florida that is disjoint with the hurricane, and vice versa. These three regions are referred to as *topological primitives*, from which the original regions can be reconstructed. Although maps were integrated at a deep level in this system, a map type was not available to the user, except through collections of the geometric primitives and visualizations. Furthermore, the map storage was utilized to provide a spatial algebra for points, lines, and regions, but not for maps. Therefore, no operational closure was provided over maps. However, this system used the map storage to enforce topological constraints over regions at the data level.

1.4 A Map Framework

An examination of maps under various contexts confirms the driving characteristics of maps, namely that maps fundamentally reflect:

- Representation of space.
- Representation of topology.
- Representation of thematic values.

The wide scale implementation of maps in spatial and geographic information systems further highlights the utility of maps as data, computational, and visualization tools. Furthermore, the need for operations over maps is highlighted in their use. For example, the proposal of cognitive maps in robotics provides a scenario in which maps of various locales may need to be combined for decision making. In a setting where various agents share maps, operations to support querying maps as fundamental data objects are required. Ideally, maps can be integrated throughout an information system as both fundamental objects which can be individually stored, processed, and queried, and high-level visualization objects. The use of maps as visualizations is mature, but the use of maps as fundamental data objects is lacking.

In this book, we propose a complete *Map Framework* that defines data types for maps, and provides operations and predicates with type closure guarantees over the map types. We construct maps by specifying an abstract model of maps, creating a discrete model of maps that preserves the properties of the abstract model, and finally, providing an implementation model of maps for database systems. The result is a complete and robust algebra that provides a fundamental data type of maps in computational systems.

References

1. Acredolo, L.P.: Developmental changes in the ability to coordinate perspectives of a large-scale space. Developmental Psychology **13**(1), 1 (1977)
2. Baumgart, B.G.: Winged Edge Polyhedron Representation. Tech. rep., Stanford University, Stanford, CA, USA (1972)
3. Baumgart, B.G.: A polyhedron representation for computer vision. In: Proceedings of the May 19-22, 1975, national computer conference and exposition, pp. 589–596. ACM (1975)
4. Billinghurst, M., Weghorst, S.: The use of sketch maps to measure cognitive maps of virtual environments. In: Virtual Reality Annual International Symposium, 1995. Proceedings., pp. 40–47. IEEE (1995)
5. Blades, M.: The reliability of data collected from sketch maps. Journal of Environmental Psychology **10**(4), 327–339 (1990)
6. Broome, F.R., Meixler, D.B.: The TIGER Data Base Structure. Cartography and Geographic Information Systems **17**(1), 39–48 (1990)
7. Downs, R.M., Stea, D.: Image and environment: Cognitive mapping and spatial behavior. Transaction Publishers (1973)
8. Estrada, C., Neira, J., Tardós, J.D.: Hierarchical slam: Real-time accurate mapping of large environments. Robotics, IEEE Transactions on **21**(4), 588–596 (2005)
9. Filho, W.C., de Figueiredo, L.H., Gattass, M., Carvalho, P.C.: A Topological Data Structure for Hierarchical Planar Subdivisions. In: 4th SIAM Conference on Geometric Design (1995)
10. Fisher, H.: Symap. In: Projects: 1960-1970. Laboratory for Computer Graphics and Spatial Analysis. (1966)
11. Guibas, L., Stolfi, J.: Primitives for the manipulation of general subdivisions and the computation of voronoi. ACM transactions on graphics (TOG) **4**(2), 74–123 (1985)
12. Güting, R.H.: Geo-Relational Algebra: A Model and Query Language for Geometric Database Systems. In: Extending Database Technology, pp. 506–527 (1988)
13. Güting, R.H., Schneider, M.: Realm-based spatial data types: the rose algebra. The VLDB JournalThe International Journal on Very Large Data Bases **4**(2), 243–286 (1995)
14. Herring, J.: TIGRIS: Topologically Integrated Geographic Information Systems. In: 8th International Symposium on Computer Assisted Cartography, pp. 282–291 (1987)
15. Herring, J.: TIGRIS: A Data Model for an Object-Oriented Geographic Information System. Computers and Geosciences **18**(4), 443–452 (1992)
16. Huang, Z., Svensson, P., Hauska, H.: Solving Spatial Analysis Problems with GeoSAL, A Spatial Query Language. In: Proceedings of the 6th Int. Working Conf. on Scientific and Statistical Database Management, pp. 1–17. Institut f. Wissenschaftliches Rechnen Eidgenoessische Technische Hochschule Zürich (1992)
17. Jefferies, M.E., Yeap, W.K.: The utility of global representations in a cognitive map. In: Spatial Information Theory, pp. 233–246. Springer (2001)
18. Kaplan, S.: Cognitive maps, human needs and the designed environment 5.4. Environmental Design Research: Selected papers **1**, 275 (1973)
19. Kaplan, S.: Cognitive maps in perception and thought. Image and environment pp. 63–78 (1973)
20. Kitchin, R.M.: Cognitive maps: What are they and why study them? Journal of environmental psychology **14**(1), 1–19 (1994)
21. Kuipers, B.: Modeling spatial knowledge. Cognitive science **2**(2), 129–153 (1978)
22. Kuipers, B.: The "map in the head" metaphor. Environment and Behavior **14**(2), 202–220 (1982)
23. Kuipers, B.: The cognitive map: Could it have been any other way? In: Spatial orientation, pp. 345–359. Springer (1983)
24. Kuipers, B.: The spatial semantic hierarchy. Artificial intelligence **119**(1), 191–233 (2000)
25. Kuipers, B., Byun, Y.T.: A robot exploration and mapping strategy based on a semantic hierarchy of spatial representations. Robotics and autonomous systems **8**(1), 47–63 (1991)

26. Ledoux, H., Gold, C.: A voronoi-based map algebra. Progress in Spatial Data Handling pp. 117–131 (2006)
27. Lynch, K.: The image of the city, vol. 11. MIT press (1960)
28. Mackworth, A.K.: On reading sketch maps. Department of Computer Science, University of British Columbia (1977)
29. Mennis, J.: Multidimensional map algebra: Design and implementation of a spatio-temporal gis processing language. Transactions in GIS **14**(1), 1–21 (2010)
30. Muller, D.E., Preparata, F.P.: Finding the intersection of two convex polyhedra. Theoretical Computer Science **7**(2), 217–236 (1978)
31. Scholl, M., Voisard, A.: Thematic Map Modeling. In: SSD '90: Proceedings of the first symposium on Design and implementation of large spatial databases, pp. 167–190. Springer-Verlag New York, Inc., New York, NY, USA (1990)
32. Steinitz, C., Parker, P., Jordan, L.: Hand-drawn overlays: Their history and prospective uses. Landscape architecture **66**(5), 444–455 (1976)
33. Tolman, E.C.: Cognitive maps in rats and men. Psychological review **55**(4), 189 (1948)
34. Tomlin, C.D.: A map algebra. Harvard Graduate School of Design (1990)
35. Tomlin, C.D.: Map algebra: one perspective. Landscape and Urban Planning **30**(1), 3–12 (1994)
36. Voisard, A., David, B.: Mapping Conceptual Geographic Models onto DBMS Data Models. Tech. Rep. TR-97-005, International Computer Science Institute (1997)
37. Wood, D., Beck, R.: Tour personality: The interdependence of environmental orientation and interpersonal behavior. Journal of Environmental Psychology **10**(3), 177–207 (1990)
38. Yeap, W.K.: Towards a computational theory of cognitive maps. Artificial Intelligence **34**(3), 297–360 (1988)

Chapter 2
A Formal Model of Maps as a Fundamental Type

The algebra defining the traditional spatial data types, their associated operations, and topological predicates form the the foundation of spatial data processing in spatial processing systems. This algebra also forms the foundation of the Map Framework, as presented in this book.

2.1 Spatial Data Models

We distinguish two *generations* of spatial data types. The types of the first generation have a simple structure, and are known as the *simple spatial types* [2, 7, 9, 11]. The simple spatial types define three structures: *points*, *lines*, and *regions*. A *simple point* describes an element of the Euclidean plane $\mathbb{R}^2$. A *simple line* is a one-dimensional, continuous geometric structure embedded in $\mathbb{R}^2$ with two end points. A *simple region* is a two-dimensional point set in $\mathbb{R}^2$ and is topologically equivalent to a closed disk. Figure 2.1 shows examples of the simple spatial types.

The simple spatial types allow for the representation of zero dimensional, one dimensional, and two dimensional structures in geographic reality, but lack two key properties. First, many objects in geographic reality have multiple *components*, but are intuitively a single object. For example, one may wish to store the shape of the country of Italy in a spatial system; Italy contains islands. A simple region cannot represent Italy's mainland and islands as a single data object. Second, the simple types do not provide type closure guarantees. For example, the intersection of two simple regions may result in multiple disconnected components, as shown in Fig. 2.2; thus, the intersection of two simple regions is not necessarily a simple region. Achieving type closure is a fundamental goal of the Map Framework, precisely to avoid such problems.

M. McKenney and M. Schneider, *Map Framework*,
DOI 10.1007/978-3-319-46766-5_2

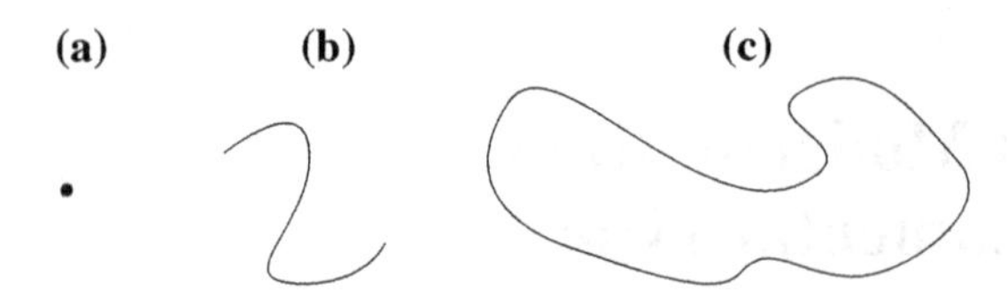

Fig. 2.1 A simple point (**a**), a simple line (**b**), and a simple region (**c**)

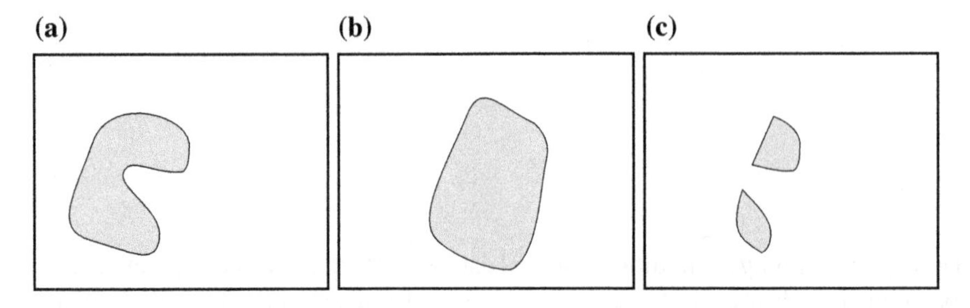

Fig. 2.2 Two simple regions (**a** and **b**), and the result of their intersection (**c**). The result is not a single simple region

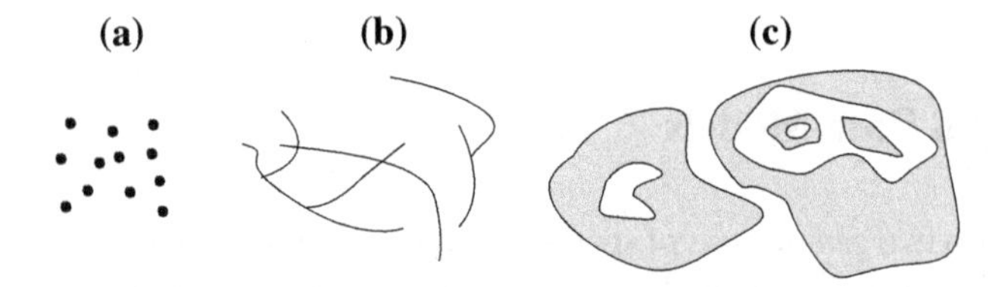

Fig. 2.3 A complex point (**a**), a complex line (**b**), and a complex region (**c**)

Additional requirements of applications as well as needed closure properties of operations led to the second generation of *complex spatial data types* [2, 21, 23] illustrated in Fig. 2.3 ([20] for a survey).

A *complex point* is a finite point collection (e.g., the positions of all lighthouses in Florida). A *complex line* is an arbitrary, finite collection of one-dimensional curves, i.e., a spatially embedded network possibly consisting of several disjoint connected components (e.g., the Nile Delta). A *complex region* permits multiple areal components, called *faces*, and holes in faces (e.g., Italy with its mainland and offshore islands as components and with the Vatican as a hole).

Two additional spatial data types for regions have been proposed as intermediate steps between simple and complex regions. These are, *composite regions* [3] and *simple regions with holes* [2, 6, 23]. A composite region can contain multiple faces, but no holes. In other words, a composite region consists of finitely many simple regions that are either disjoint, or meet at single points. Although holes are not allowed in composite regions, "hole-like" configurations can exist if two components of one region touch at a single point of their boundaries at two different locations (Fig. 2.4).

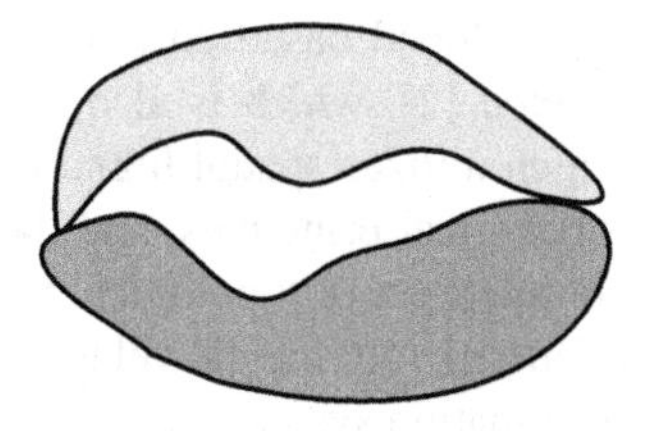

Fig. 2.4 Sample composite region with two components presenting a hole-like structure

A simple region with holes contains only a single face, with finitely many holes. The holes in a simple region with holes are allowed to meet at a point, but cannot form a configuration that causes the interior of the region to be disconnected. In other words, a hole-like structure cannot be formed by the holes in a simple region with holes. Intuitively, a simple region with holes is a complex region that has only a single face.

Each spatial type is defined such that it is made up of three parts: the *interior*, *boundary*, and *exterior*. Given a spatial object A, these components are indicated respectively as A°, ∂A, and A^-. For example, the boundary of a line is its endpoints, and its interior consists of the lines that connect the endpoints. The exterior of a line consists of all points in $\mathbb{R}^2$ that are not part of the interior or boundary. Similarly, the boundary of a region is the line that defines the region's border. The interior consists of all points that lie inside the region, and the exterior consists of all points that are not part of the boundary or interior. These concepts are required for the definition of topological relationships between spatial types (defined in the next section).

2.2 Topological Relationships

In the development of the Map Framework, we rely heavily on concepts of topology and topological relationships between spatial data types. The study of topological relationships between objects in space has been the subject of a vast amount of research [1–4, 6, 8, 9, 14–19, 21]. In the areas of databases and GIS, the motivation for formally defining topological relationships between spatial objects has been driven by a need for querying mechanisms that can filter these objects in spatial selections or spatial joins based on their topological relationships to each other, as well as a need for appropriate mechanisms to aid in spatial data analysis and spatial constraint specification.

Topological relationships indicate qualitative properties of the relative positions of spatial objects that are preserved under continuous transformations such as translation, rotation, and scaling. Quantitative measures such as distance or size measurements are deliberately excluded in favor of modeling notions such as connectivity, adjacency, disjointedness, inclusion, and exclusion. Attempts to model and rigorously define the relationships between certain types of spatial objects have lead to the development of three popular approaches: the *9-intersection model* [9], which is developed based on point set theory and point set topology, the *calculus based method* [2], which is also based on point set topology, and the *RCC model* [19], which utilizes spatial logic. Because the definitions of spatial objects are based on topological principles, and the inability of the calculus based method to identify a complete set of topological relationships, the 9-intersection model is typically used to model topological relationships between spatial objects in the field of spatial information systems.

The 9-intersection model of topological relationships characterizes the topological relationship between two spatial objects by evaluating the non-emptiness of the intersection between all combinations of the interior ($^\circ$), boundary (∂) and exterior ($^-$) of the objects involved. A unique 3×3 matrix, termed the *9-intersection matrix* (9IM), with values filled as illustrated in Fig. 2.5 describes the topological relationships between a pair of objects.

The 9IM is general in the sense that it is applicable to any spatial data types that maintain the topological concepts of interior, exterior, and boundary. Applying the model to different data types results in different sets of applicable topological relationships. For example, Fig. 2.6 depicts the 8 topological relationships between simple regions. There are 33 topological relationships between complex regions [21]. In the construction of a data type for maps, we will primarily refer to the topological relationships *disjoint* and *meet* between pairs of simple regions and pairs of complex regions. The 9IM for those relationships is identical for both simple and complex regions.

Various models of topological predicates based on the 9-intersection model using both *component derivations*, in which relationships are derived based on the interactions of all components of spatial objects, and *composite derivations*, in which relationships model the global interaction of two objects, exist in the literature. Examples

$$\begin{pmatrix} A^\circ \cap B^\circ \neq \emptyset & A^\circ \cap \partial B \neq \emptyset & A^\circ \cap B^- \neq \emptyset \\ \partial A \cap B^\circ \neq \emptyset & \partial A \cap \partial B \neq \emptyset & \partial A \cap B^- \neq \emptyset \\ A^- \cap B^\circ \neq \emptyset & A^- \cap \partial B \neq \emptyset & A^- \cap B^- \neq \emptyset \end{pmatrix}$$

Fig. 2.5 The 9-intersection matrix for spatial objects A and B

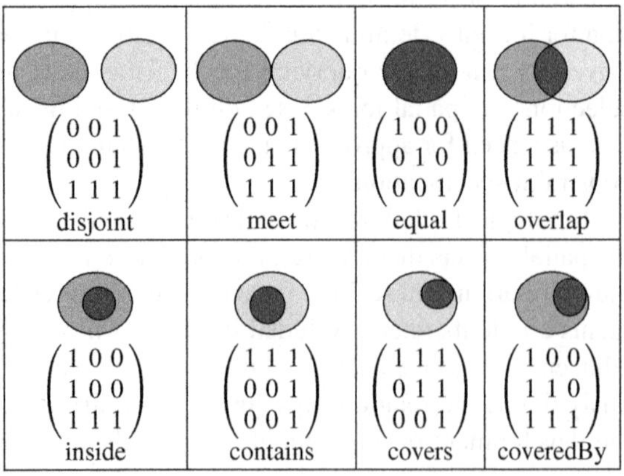

$\begin{pmatrix} 0&0&1 \\ 0&0&1 \\ 1&1&1 \end{pmatrix}$	$\begin{pmatrix} 0&0&1 \\ 0&1&1 \\ 1&1&1 \end{pmatrix}$	$\begin{pmatrix} 1&0&0 \\ 0&1&0 \\ 0&0&1 \end{pmatrix}$	$\begin{pmatrix} 1&1&1 \\ 1&1&1 \\ 1&1&1 \end{pmatrix}$
disjoint	meet	equal	overlap
$\begin{pmatrix} 1&0&0 \\ 1&0&0 \\ 1&1&1 \end{pmatrix}$	$\begin{pmatrix} 1&1&1 \\ 0&0&1 \\ 0&0&1 \end{pmatrix}$	$\begin{pmatrix} 1&1&1 \\ 0&1&1 \\ 0&0&1 \end{pmatrix}$	$\begin{pmatrix} 1&0&0 \\ 1&1&0 \\ 1&1&1 \end{pmatrix}$
inside	contains	covers	coveredBy

Fig. 2.6 The 8 topological predicates between simple regions. One object is *shaded dark* and the other *light* as in the *disjoint* relationship, whereas the shared areas have the *darkest shade*

of component derivations can be found in [3, 6]. In [6], the authors define topological relationships between regions with holes in which each of the relationships between all faces and holes are calculated. Given two regions, R and S, containing m and n holes respectively, a total of $(n + m + 2)^2$ topological predicates are possible. It is shown that this number can be reduced $mn + m + n + 1$; however, the total number of predicates between two objects depends on the number of holes the objects contain. Similarly, in [3], predicates between complex regions without holes are defined based on the topological relationship of each face within one region with all other faces of the same region, all faces of the other region, and the entire complex regions themselves. Given regions S and R with m and n faces respectively, a matrix is constructed with $(m + n + 2)^2$ entries that represent the topological relationships between S and R and each of their faces.

The most basic example of a composite derivation model (in which the global interaction of two spatial objects is modeled) is the derivation of topological predicates between simple spatial objects in [9]. This model has been used as the basis for modeling topological relationships between object components in the component models discussed above. In [21], the authors apply an extended 9-intersection model to point sets belonging to complex points, lines, and regions. Based on this application, the authors are able to construct a composite derivation model for complex data types and derive a complete and finite set of topological predicates between them, thus resolving the main drawback of the component derivation model.

More recently, it has been observed that composite models of topological relationships between spatial objects are *global*, in that they characterize an entire scene by a single topological relationship that may hide *local* information about the object's relationship [17]. The hiding of local information is expressed in two ways in global topological relationship models: through the *dominance problem*, and the *composition problem*. The dominance problem indicates the property that the global view exhibits *dominance* properties among the topological relationships as defined by the 9-intersection model. For example, while building roads between two adjacent countries, one might be interested to know that there is a disjoint island in one of the countries for which a bridge to the other country is required. The disjointedness in this case is overshadowed or dominated by the existing *meet* (adjacent) situation between the countries' mainlands. The composition problem expresses the property that a global topological predicate may indicate a certain relationship between two objects that does not exist locally. For example, consider two complex regions that have individual faces that satisfy the *inside*, *covers*, and *meet* predicates. Globally, this configuration satisfies the overlap predicate even though no faces overlap locally. These properties have been addressed through the *local topological relationship* models between composite regions [17] and between complex regions [18]. These approaches model a topological relationship between two multicomponent objects based on the topological relationships that exist between the components of the objects. Furthermore, it is shown that these models are more expressive than the global 9IM models in that they can distinguish all the topological scenes that the 9IM models can distinguish, plus many more.

2.3 An Informal Overview of Spatial Partitions

In this paper, we model maps as *spatial partitions*, as discussed in [10, 12–14]. The definition of spatial partitions is rather dense, so we begin by providing an intuitive description of them, and then present the formal definition in later sections.

A spatial partition, in two dimensions, is a subdivision of the plane into pairwise disjoint *regions* such that each region is associated with a *label* or *attribute* having simple or complex structure, and these regions are separated from each other by *boundaries*. The label of a region describes the thematic data associated with the region. All points within the spatial partition that have an identical label are part of the same region. Topological relationships are implicitly modeled among the regions in a spatial partition. For instance, neglecting common boundaries, the regions of a partition are always disjoint; this property causes maps to have a rather simple structure. Note that the *exterior* of a spatial partition (i.e., the unbounded face) is always labeled with the $\perp$ symbol. Figure 2.7a depicts an example spatial partition consisting of two regions.

We stated above that each region in a spatial partition is associated with a single attribute or label. A spatial partition is modeled by mapping Euclidean space to such labels. Labels themselves are modeled as sets of attributes. The regions of the spatial partition are then defined as consisting of all points which contain an identical label. Adjacent regions each have different labels in their interior, but their common boundary is assigned the label containing the labels of both adjacent regions. Figure 2.7b shows an example spatial partition complete with boundary labels.

In [10], operations over spatial partitions are defined based on known map operations in the literature. It is shown that all known operations over spatial partitions can be expressed in terms of three fundamental operations: intersection, relabel, and refine. Furthermore, the type of spatial partitions is shown to be closed under these operations, indicating that the type of spatial partitions is closed under all known operations over them. We will cover operations over spatial partitions in later chapters.

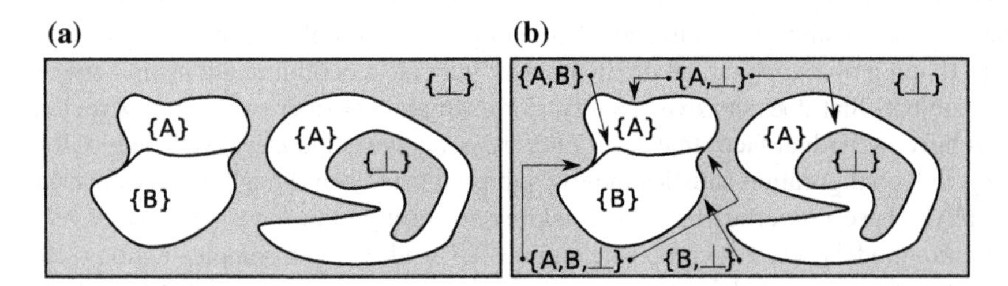

Fig. 2.7 A sample spatial partition with two regions. **a** The spatial partition annotated with region labels. **b** The spatial partition with its region and boundary labels. Note that labels are modeled as sets of attributes in spatial partitions

2.4 Spatial Partitions: A Mathematical Model of Maps

Spatial partitions have been developed in the literature in [10, 12]; those definitions require some modification in order to define both topological predicates over maps and the discrete model of spatial partitions suitable for implementation. We present the modified definition here. We first introduce the mathematical notation and definitions required to formally define spatial partitions. Then, the formal mathematical type definition of spatial partitions is presented.

2.4.1 Notation

The application of a function $f : A \to B$ to a set of values $S \subseteq A$ is defined as:

$$f(S) := \{f(x) | x \in S\} \subseteq B \tag{2.1}$$

In some cases we know that $f(S)$ returns a singleton set, in which case we write $f[S]$ to denote the single element:

$$f(S) = \{y\} \Leftrightarrow f[S] = y \tag{2.2}$$

The inverse function $f^{-1} : B \to 2^A$ of f is defined as:

$$f^{-1}(y) := \{x \in A | f(x) = y\} \tag{2.3}$$

It is important to note that f^{-1} is a total function and that f^{-1} applied to a set yields a set of sets.

We define the range function of a function $f : A \to B$ that returns the set of all elements that f returns for an input set A as:

$$rng(f) := f(A) \tag{2.4}$$

Let (X, U) be a topological space [5] with topology $U \subseteq 2^x$ defined by the basis β (a basis of U is a set such that every element of U is equivalent to a union of elements in the basis). Let $S \in \beta$. The *interior* of S, denoted by S°, is defined as the union of all open sets that are contained in S. The *closure* of S, denoted by $\overline{S}$ is defined as the intersection of all closed sets that contain S. The *exterior* of S is given by $S^- := (X - S)^\circ$, and the *boundary* or *frontier* of S is defined as $\partial S := \overline{S} \cap \overline{X - S}$. An open set is *regular* if $A = \overline{A}^\circ$ [22]. In this paper, we deal with the topological space $\mathbb{R}^2$, i.e., $X = \mathbb{R}^2$.

A *partition* of a set S, requires U to be a *partition topology*. A partition topology, is a topology whose basis β is complete decomposition into non-empty, disjoint subsets $\{b_i | i \in I\}$, called *blocks*, where I is an index set used to name different blocks, where:

$$\begin{aligned}&(\text{i})\ \forall i \in I : b_i \neq \emptyset,\\&(\text{ii})\ \bigcup_{i\in I} b_i = \beta, \text{ and}\\&(\text{iii})\ \forall i, j \in I, i \neq j : b_i \cap b_j = \emptyset,\end{aligned} \tag{2.5}$$

We can equivalently represent a partition as a total and surjective function:

$$f : \beta \rightarrow I \tag{2.6}$$

such that the above constraints are satisfied. However, a spatial partition cannot be defined simply as a set-theoretic partition of the plane, that is, as a partition of $\mathbb{R}^2$ or as a function $f : \mathbb{R}^2 \rightarrow I$, for two reasons: first, f cannot be assumed to be total in general, and second, f cannot be uniquely defined on the borders between adjacent subsets of $\mathbb{R}^2$. In the following, we build upon the concept of partitions to create spatial partitions.

2.4.2 Spatial Partitions

In this section, we build the definition of spatial partitions based upon the notation developed for partitions, as in [10]. First, we define a *spatial mapping*, then impose constraints on a spatial mapping to create a spatial partition.

A *spatial mapping* of type A is a total function $\pi : \mathbb{R}^2 \rightarrow 2^A$ that maps points in the plane to an element of 2^A where A is a set of *labels*. In this context, a label is a general term and no restrictions are imposed on labels; typically, applications will use labels to represent thematic values or identifiers. The existence of an undefined element $\perp_A$ is required to represent undefined labels (i.e., the exterior of a partition).

Although there are not restrictions on a spatial mapping, the mapping will result in sets of points that map to the same element of 2^A; we denote such sets of points as *components* of the partition imposed by the spatial mapping. Definition 2.1 identifies the different components of a partition imposed by a spatial mapping.

A region in the partition imposed by a spatial mapping (Definition 2.1(i)) is a point set that maps the same label, and that label is an element of the set A. It is important to note that a region in a spatial partition is intuitively the same as a complex region; thus, the region is not necessarily connected (may contain multiple *faces*), and may contain holes. A region in a spatial mapping requires the further constraints imposed by a spatial partition to ensure that regions are mathematically equivalent to complex regions. We use the term *regions* to refer to complex regions in the remainder of this book. The *border* (Definition 2.1(ii)) between two regions will have the labels of both neighboring regions, and thus, will be labeled with an element of 2^A with cardinality greater than 1.

The *interior* of spatial mapping π (Definition 2.1(iii)) is defined as the union of π's regions. Thus, the interior of the spatial mapping is identified by points in the plane that map to a single set that exists in 2^A. The union of all borders in a spatial mapping π form the *boundary* of π (Definition 2.1(iv)). Therefore, the boundary of the of the partition consists of all points that are labeled with non-singleton sets of 2^A. The *exterior* of π (Definition 2.1(v)) is the point set such that each point in that set maps to the label $\perp_A$. The exterior behaves similarly to the exterior of a complex region. The exterior may have multiple connected components. The unique feature about the exterior is that is an *unbounded face*.

Definition 2.1 Let π be a spatial mapping of type A

(i) $\rho(\pi) := \pi^{-1}(rng(\pi) \cap \{X \in 2^A | \, |X| = 1\})$ (*regions*)

(ii) $\omega(\pi) := \pi^{-1}(rng(\pi) \cap \{X \in 2^A | \, |X| > 1\})$ (*borders*)

(iii) $\pi^\circ := \bigcup_{r \in \rho(\pi) | \pi[r] \neq \{\perp_A\}} r$ (*interior*)

(iv) $\partial\pi := \bigcup_{b \in \omega(\pi)} b$ (*boundary*)

(v) $\pi^- := \pi^{-1}(\{\perp_A\})$ (*exterior*)

As an example, let π be the spatial partition in Fig. 2.7 of type $X = \{A, B, \perp\}$. In this case, $rng(\pi) = \{\{A\}, \{B\}, \{\perp\}, \{A, B\}, \{A, \perp\}, \{B, \perp\}, \{A, B, \perp\}\}$. Therefore, the regions of π are the blocks labeled $\{A\}$, $\{B\}$, and $\{\perp\}$ and the boundaries are the blocks labeled $\{A, B\}$, $\{A, \perp\}$, $\{B, \perp\}$, and $\{A, B, \perp\}$.

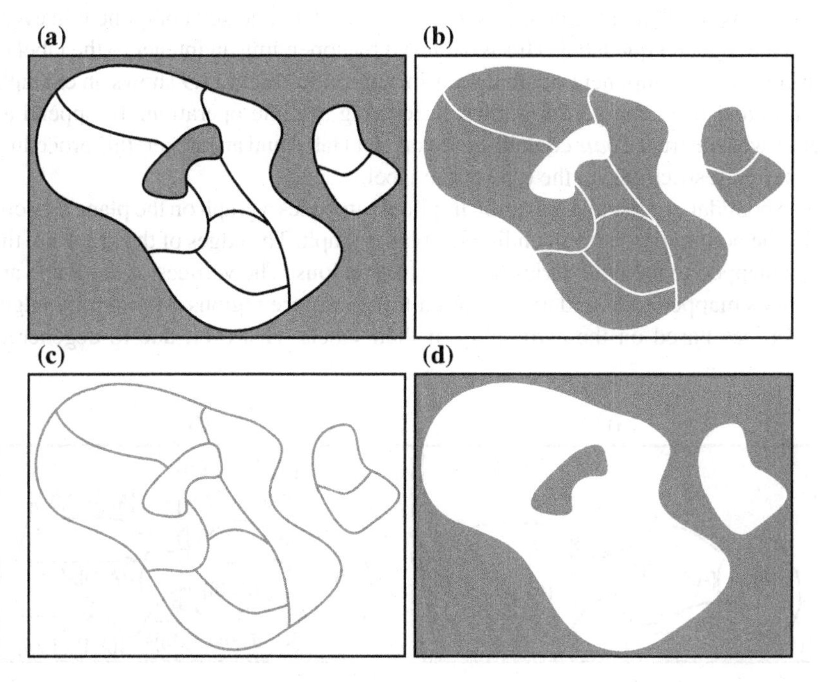

Fig. 2.8 **a** A spatial partition π with two disconnected faces, one containing a hole. **b** The interior (π°). **c** The boundary ($\partial\pi$). **d** The exterior (π^-). Note that the labels have been omitted in order to emphasize the components of the spatial partition

Figure 2.8 provides a pictorial example of the interior, exterior, and boundary of an example spatial mapping (note that the borders and boundary consist of the same points, but the boundary is a single point set whereas the borders are a set of point sets).

A *spatial partition* of type A is then defined as a spatial mapping of type A whose regions are regular open sets [22] and whose borders are labeled with the union of labels of all adjacent regions. From this point forward, we use the term *partition* to refer to a spatial partition.

Definition 2.2 A spatial partition of type A is a spatial mapping π of type A with:
(i) $\forall r \in \rho(\pi) : r = \overline{r}^{\circ}$
(ii) $\forall b \in \omega(\pi) : \pi[b] = \{\pi[[r]] | r \in \rho(\pi) \wedge b \subseteq \partial r\}$

The remaining portion of the definition of spatial partitions requires the use of the *refine operation* over spatial partitions. This operation is formally defined in Chap. 3, so we provide an intuitive definition here.

The refine operation over spatial partitions uniquely identifies the connected components of a partition. Recall that two regions in a partition can share the same label if they are disjoint or meet at discrete points along their boundaries. Given a partition π containing regions with multiple connected components (recall that each connected component of a region has the same label), the operation *refine*(π) returns a partition with identical structure to π, but with each connected component in every region having a unique label. This is achieved by appending an integer to the label of each connected component of all regions in a partition. Figure 2.9 shows an example partition and the same partition after performing a refine operation. To append an integer, we construct a *pair* containing the original label and an integer; this procedure imposes no restrictions on the type of the label.

The boundary of a spatial partition implicitly imposes a graph on the plane. Specifically, the boundaries form an undirected planar graph. The edges of the graph are the points mapped to the boundaries between two regions. The vertices of the graph are the points mapped to boundaries between three or more regions. We identify edges and vertices based on the cardinality of their labels. However, due to degenerate

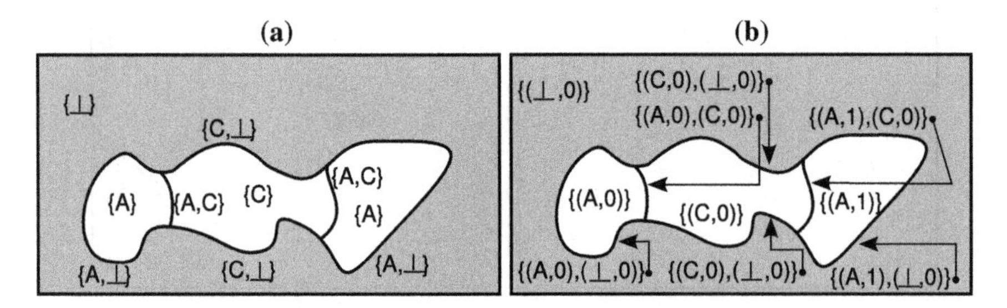

Fig. 2.9 The application of the refine operation to a spatial partition. A spatial partition with two regions and its boundary and region labels (**a**). The result of the refine operation on (**a**). The region labels are $\{A\}$ and $\{C\}$ in (**a**) and $\{(A, 0)\}$, $\{(C, 0)\}$, and $\{(A, 1)\}$ in (**b**). The exterior label is $\perp$ in (**a**) and $(\perp, 0)$ in (**b**). All other labels are boundaries

cases, we must use the refinement of a partition to identify these features. We define the set of edges and vertices imposed on the plane by a spatial partition as follows:

Definition 2.3 A boundary point of a spatial partition π is classified as being a *vertex* or as being part of an *edge* by examining the refinement $\sigma = \mathit{refine}(\pi)$ as follows:
(i) $\varepsilon(\pi) = \{b \in \omega : |\sigma[b]| = 2\}$
(ii) $\nu(\pi) = \{b \in \omega : |\sigma[b]| > 2\}$

2.5 Summary

Spatial partitions provide a data type to represent map constructs and their associated attribute values. The power of the spatial partition data type becomes clear in later sections where we show that nearly all operations over spatial partitions can be expressed as compositions of three fundamental *power operations*. We will prove type closure of those operations under the type of spatial partitions such that any operation defined as a composition of the power operations inherits the type closure properties. Furthermore, complex regions are a special case of spatial partitions which makes spatial partitions a useful tool for studying regions and their operations, in addition to maps and their associated operations.

References

1. Clementini, E., Di Felice, P.: A comparison of methods for representing topological relationships. Information Sciences-Applications **3**(3), 149–178 (1995)
2. Clementini, E., Di Felice, P.: A Model for Representing Topological Relationships between Complex Geometric Features in Spatial Databases. Information Systems **90**, 121–136 (1996)
3. Clementini, E., Di Felice, P., Califano, G.: Composite Regions in Topological Queries. Information Systems **20**, 579–594 (1995)
4. Clementini, E., Di Felice, P., Van Oosterom, P.: A small set of formal topological relationships suitable for end-user interaction. In: Advances in Spatial Databases, pp. 277–295. Springer (1993)
5. Dugundi, J.: Topology. Allyn and Bacon (1966)
6. Egenhofer, M.J., Clementini, E., Di Felice, P.: Topological Relations between Regions with Holes. Int. Journal of Geographical Information Systems **8**, 128–142 (1994)
7. Egenhofer, M.J., Frank, A., Jackson, J.P.: A Topological Data Model for Spatial Databases. In: 1st Int. Symp. on the Design and Implementation of Large Spatial Databases, pp. 271–286. Springer-Verlag (1989)
8. Egenhofer, M.J., Franzosa, R.D.: Point-set topological spatial relations. International Journal of Geographical Information System **5**(2), 161–174 (1991)
9. Egenhofer, M.J., Herring, J.: Categorizing Binary Topological Relations Between Regions, Lines, and Points in Geographic Databases. Technical report, National Center for Geographic Information and Analysis, University of California, Santa Barbara (1990)
10. Erwig, M., Schneider, M.: Partition and Conquer. In: 3rd Int. Conf. on Spatial Information Theory (COSIT), pp. 389–408. Springer-Verlag (1997)
11. Frank, A., Kuhn, W.: Cell Graphs: A Provable Correct Method for the Storage of Geometry. In: 2rd Int. Symp. on Spatial Data Handling, pp. 411–436 (1986)

12. McKenney, M., Schneider, M.: Advanced Operations for Maps in Spatial Databases. In: Int. Symp. on Spatial Data Handling (2006)
13. McKenney, M., Schneider, M.: Spatial Partition Graphs: A Graph Theoretic Model of Maps. In: Int. Symp. on Spatial and Temporal Databases (SSDT), pp. 167–184 (2007)
14. McKenney, M., Schneider, M.: Topological Relationships Between Map Geometries. In: Advances in Databases: Concepts, Systems and Applications, 13th International Conference on Database Systems for Advanced Applications (2007)
15. McKenney, M., Praing, R., Schneider, M.: Deriving topological relationships between simple regions with holes. In: Headway in Spatial Data Handling, pp. 521–531. Springer (2008)
16. McKenney, M., Pauly, A., Praing, R., Schneider, M.: Dimension-Refined Topological Predicates. In: 13th ACM International Symposium on Geographic Information Systems, pp. 240–249 (2005)
17. McKenney, M., Pauly, A., Praing, R., Schneider, M.: Preserving Local Topological Relationships. In: ACM Symp. on Geographic Information Systems (ACM GIS), pp. 123–130. ACM (2006)
18. McKenney, M., Pauly, A., Praing, R., Schneider, M.: Local Topological Relationships for Complex Regions. In: Int. Symp. on Spatial and Temporal Databases (SSDT), pp. 203–220 (2007)
19. Randell, D.A., Cui, Z., Cohn, A.: A Spatial Logic Based on Regions and Connection. In: International Conference on Principles of Knowledge Representation and Reasoning, pp. 165–176 (1992)
20. Schneider, M.: Spatial Data Types for Database Systems - Finite Resolution Geometry for Geographic Information Systems, vol. LNCS 1288. Springer-Verlag, Berlin Heidelberg (1997)
21. Schneider, M., Behr, T.: Topological Relationships between Complex Spatial Objects. ACM Trans. on Database Systems (TODS) **31**(1), 39–81 (2006)
22. Tilove, R.B.: Set Membership Classification: A Unified Approach to Geometric Intersection Problems. IEEE Trans. on Computers **C-29**, 874–883 (1980)
23. Worboys, M.F., Bofakos, P.: A Canonical Model for a Class of Areal Spatial Objects. In: 3rd Int. Symp. on Advances in Spatial Databases, pp. 36–52. Springer-Verlag (1993)

Chapter 3
PLR Partitions: Extending Maps to Include Point and Line Features

The model of spatial partitions provides a data type that allows a user to represent maps containing *regions* such that each region may contain thematic information in a *label*. However, spatial partitions cannot represent point and line geometries apart from the points and lines at which regions meet in a partition. In this chapter, we define Point, Line, Region (PLR) Partitions that can represent maps containing point, line, and region features.

3.1 Spatial Data Models

Chapter 2 opened with a discussion of spatial data models. Here, we review relevant spatial models with an emphasis on the concepts more relevant to PLR Partitions.

Spatial partitions provide a data model for maps that include *region* features. Specifically spatial partitions include complex regions such that each region contains a label. Indeed, a spatial partition with with type A such that $|A| = 2$ is equivalent to a complex region (in other words, A contains a single region label and the exterior label). The goal of PLR partitions is to include complex regions, complex lines, and complex points together into a single map data type, and maintain the property that each component of the partition carries a label. The complex spatial types [2, 21, 28] are illustrated in Fig. 3.1 ([20] for a survey).

A *complex point* is a finite point collection (e.g., the positions of all lighthouses in Florida). A *complex line* is an arbitrary, finite collection of one-dimensional curves, i.e., a spatially embedded network possibly consisting of several disjoint connected components (e.g., the Nile Delta). A *complex region* permits multiple areal components, called *faces*, and holes in faces (e.g., Italy with its mainland and offshore islands as components and with the Vatican as a hole).

The complex types are defined based on point set topology. Specifically, each type is defined based on three parts: the *interior*, *boundary*, and *exterior*. Given a spatial object A, these components are indicated respectively as A°, ∂A, and A^-. For example, the boundary of a line is its endpoints, and its interior consists of the

M. McKenney and M. Schneider, *Map Framework*,
DOI 10.1007/978-3-319-46766-5_3

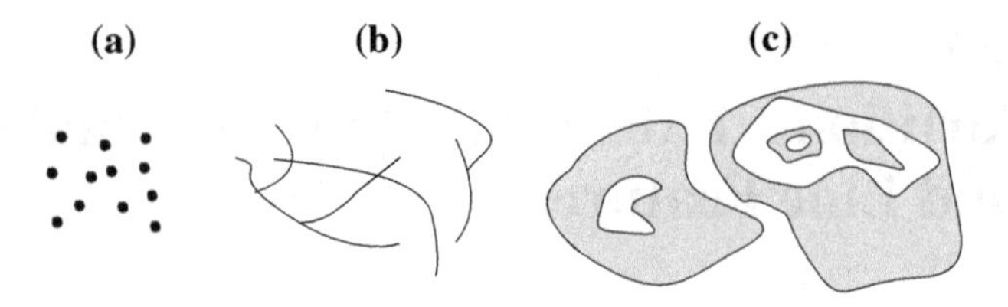

Fig. 3.1 A complex point (**a**), a complex line (**b**), and a complex region (**c**)

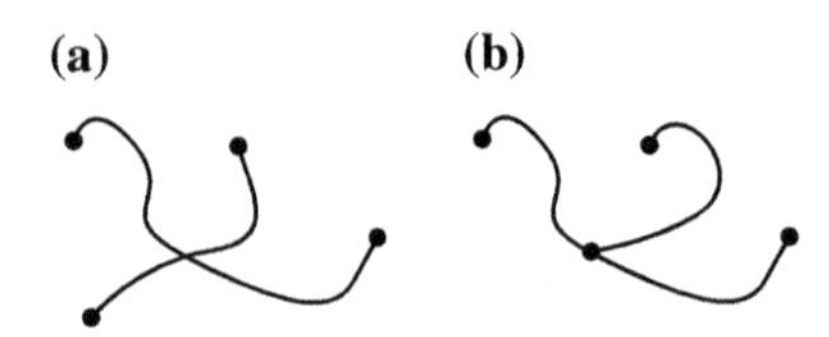

Fig. 3.2 Scenes depicting two possible topological relationships between lines. The 9IMs for these scenes are not valid for pairs of regions, and must be specifically handled in PLR partitions

lines that connect the endpoints. The exterior of a line consists of all points in $\mathbb{R}^2$ that are not part of the interior or boundary. Similarly, the boundary of a region is the line that defines the region's border. The interior consists of all points that lie inside the region, and the exterior consists of all points that are not part of the boundary or interior. The components of PLR partitions will have similar definitions. Note that a point is unique in that its interior is empty; a point has only a boundary and exterior.

The regions in a PLR partition will behave identically to the regions in a spatial partition, thus a spatial partition is a special case of a PLR partition. The discussion of topological relationships between regions is not repeated here (See Chap. 2). However, because PLR partitions contain lines and points, we must consider topological relationships between those spatial types.

The interior of a region has a non-zero area, and the boundary of a region contains its interior. Lines are different in that the interior of a line has a length, but no area, and the boundary of a line consists of its end points. Because a line's boundary does not surround, or contain, its interior, the number of topological relationships between lines is greater than the number of relationships between regions. For example, the interior of two lines may intersect without their either of their boundaries intersecting the opposing line's interior or boundary; such a situation is impossible for regions since their interiors are completely contained by their boundaries. There are 33 possible topological relationships between a pair of simple lines [3, 5, 6] and 82 possible topological relationships between a pair of complex lines [21].

The unique properties of lines lead to situations in PLR partitions that do not arise in spatial partitions. In a spatial partition, two regions always meet at region boundaries. In PLR partitions it is possible that two lines share interior points, or one line has a boundary point on the interior of a second line. These situations require special care in PLR partitions. Figure 3.2 depicts such examples.

Despite the large amount of research into thematic map computation and representation [1, 7–16, 18, 19, 22, 23, 25–27], relatively little work has looked into the union of collections of multiple types of spatial objects and thematic values into a single data type. PLR partitions, first proposed in [17], provide such a data type.

3.2 A Need for PLR Partitions

Spatial Partitions present a concept of spatial types that departs drastically from earlier conceptions. Recall that the *traditional* spatial types represent a single object in a single dimension and without an integrated notion of thematic data. For example, a point is a zero dimensional object, a line is a one dimensional object, and a region is a two dimensional object. Thus, traditional spatial models are *disjoint dimension models*, meaning that spatial objects are defined based on their dimensionality (i.e., points, lines, and regions). A common problem associated with such models arises with the use of spatial operations. For example, consider the two regions in Fig. 3.3a. In this case, the intersection of the regions does not result in a region; however, the spatial intersection between regions is typically defined such that a new region is returned. In practice, this situation is handled by defining specific intersection operations that take two regions and return a point, a line, or a region, respectively. The assumption is that the user will indicate what type of object they wish to compute by choosing the appropriate operation. However, the intersection of the regions in this case results in both a point and a line. It is clear that defining multiple intersection operations does not address the fundamental problem. In order to compute this information, the user must compute two different spatial intersection operations. The end result is that no matter which operation is chosen, information about the intersection may be lost. We denote this problem the *dimension reduction problem* for spatial operations.

A second problem that follows from the disjoint dimension model of spatial data types is that the model implicitly imposes the restriction that an object must contain features of only a single dimension. In practice, this constraint causes current spatial data models to have limitations in their abilities to represent geographic reality. For instance, consider Fig. 3.3b, c. These figures each depict an object containing components of multiple dimension: a river system containing lakes and the river feeding them (**b**), and United States Federal Government properties including the District of Columbia, the Baltimore FBI building, and the National Science Foundation building (**c**). Each of these figures contain what can intuitively be modeled as a single object containing features of multiple dimensions. These objects cannot be modeled by a single spatial object using current spatial models. For instance, assuming that the river in **b** is to be represented as a line, then either the disjoint one-dimensional sections must be grouped into an object, or a representative path of the river through the two-dimensional objects (lakes) must be arbitrarily chosen. Similarly in **c**, current spatial models will not allow a single object to represent both the two-dimensional area of the District of Columbia along with the points indicating the federal buildings, even

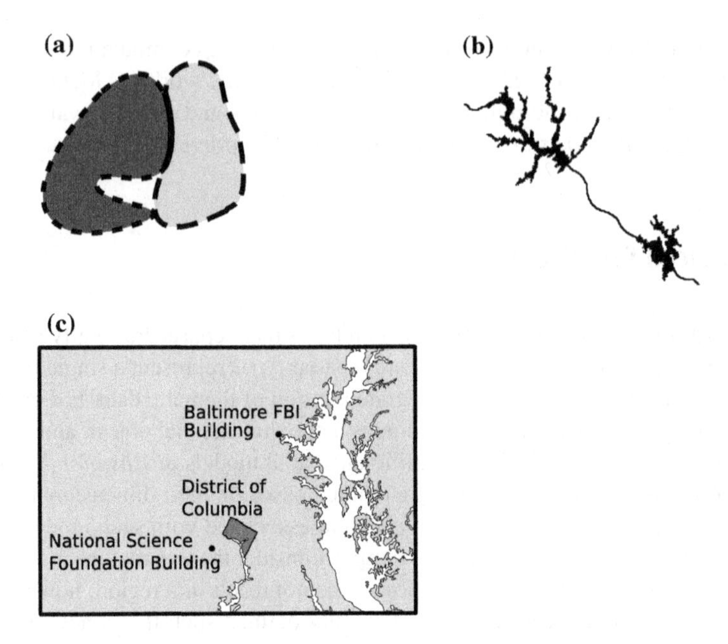

Fig. 3.3 Scenes that require a mixture of traditional types for representation

though they are all federal properties. We term this inability to model components of multiple dimensions of a single object the *dimension representation problem* of spatial object models.

Finally, traditional spatial models do not allow multiple components of an object to be identified separately from each other. For instance, if a region has two faces, both of those faces belong to the region and cannot be identified separately. Thus, two faces in a region cannot meet along a common boundary since they will be indistinguishable from their union. A consequence of this problem occurs when modeling bodies of water within a state. One can model a lake in the interior of a state as a hole in the state, even though the lake belongs to the state (the state patrols the lake, builds structures that extend into the lake, etc.). The only choices for representation are to not represent that lake in the region representing the state (which is geographically inaccurate) or to model it as a hole (which does not indicate the state's ownership of the lake). In either case, a tradeoff occurs when modeling geographic or political reality. We term this situation the *feature restriction problem*, because it arises from fundamental restrictions of traditional spatial models. A solution to this problem is to map the state and the lake as separate spatial objects, but this effectively separates fundamental information about geographic reality into separate objects that must be merged computationally for viewing.

3.3 An Informal Overview of PLR Partitions

Geometrically, a spatial partition is a subdivision of the plane into pairwise disjoint regions such that each region has a unique label. PLR partitions cannot be simply described as a subdivision of the plane because PLR partitions may contain individual, discrete points that carry a meaningful label, or individual lines with their own meaningful label that may cross regions. The ability to contain point and line structures allows PLR partitions to represent maps that more closely align with visual maps, for example, road maps with cities, roads, and states. We refer to the points, lines, and regions in a PLR partition as the *components* of the partition.

Similarly to spatial partitions, a component of a PLR partition consists of the point set containing all points that carry the same label. However, this raises two problems, in particular, in PLR partitions. The first problem is exemplified by a city that lies in a state. Often, a city will be represented as a single point, while a state will be represented as a region. The state must be defined as a regular open point set; since the point representing a city will have a different label than the rest of the state, the state will effectively have a puncture, rendering it non-regular. This is a fundamental problem that violates the basic definition of a spatial partition. Therefore, regions in a PLR partition will be modeled as they are in spatial partitions: effectively, the city must have a label that contains the city attribute information as well as the state attribute information, and the method of identifying the point set describing the state must be altered from the method used in spatial partitions. The second problem is similar, and is exemplified by lines. A line may exist in a region, in an analogous situation as the city in a state, but lines provide even more problematic situations in that they may overlap the border between two regions, or multiple borders between multiple regions. Again, we cannot simply say that a line object is identified as all points that carry the same label since that results in regions with cuts, or lines where different portions of the same line contain different labels (when a line crossed from one region to another).

The address the problems, we must define the components of PLR partitions with more explicit knowledge of the topology of components in a partition. As was mentioned previously, a point in the plane must carry the labels of all geometries that contain the point. By examining the points surrounding a particular point, we will be able determine which labels are relevant to region components, line components, and point components. Therefore, we will impose no new restrictions on labels in the sense that labels are *items* that may be implemented in any fashion, but we will impose new rules to indicate that a portion of a label (i.e., one item in the label) belongs to a particular component in the PLR partition.

As with Spatial Partitions, we will define PLR partitions using mappings of points in the plane to labels and imposing restrictions on the mappings. The fundamental notation is shared with spatial partitions, but is repeated here. Once the notation is defined, the concept of a spatial mapping for PLR partitions is developed (similar to spatial partitions), and then restrictions are imposed over the spatial mappings to define PLR partitions and their components.

3.4 Notation

We review the notation from Chap. 2. The application of a function $f : A \rightarrow B$ to a set of values $S \subseteq A$ is defined as:

$$f(S) := \{f(x)|x \in S\} \subseteq B \tag{3.1}$$

In some cases we know that $f(S)$ returns a singleton set, in which case we write $f[S]$ to denote the single element:

$$f(S) = \{y\} \Leftrightarrow f[S] = y \tag{3.2}$$

The inverse function $f^{-1} : B \rightarrow 2^A$ of f is defined as:

$$f^{-1}(y) := \{x \in A|f(x) = y\} \tag{3.3}$$

It is important to note that f^{-1} is a total function and that f^{-1} applied to a set yields a set of sets.

We define the range function of a function $f : A \rightarrow B$ that returns the set of all elements that f returns for an input set A as:

$$rng(f) := f(A) \tag{3.4}$$

Let (X, U) be a topological space [4] with topology $U \subseteq 2^x$ defined by the basis β (a basis of U is a set such that every element of U is equivalent to a union of elements in the basis). Let $S \in \beta$. The *interior* of S, denoted by S°, is defined as the union of all open sets that are contained in S. The *closure* of S, denoted by $\overline{S}$ is defined as the intersection of all closed sets that contain S. The *exterior* of S is given by $S^- := (X - S)^\circ$, and the *boundary* or *frontier* of S is defined as $\partial S := \overline{S} \cap \overline{X - S}$. An open set is *regular* if $A = \overline{A}^\circ$ [24]. In this paper, we deal with the topological space $\mathbb{R}^2$, i.e., $X = \mathbb{R}^2$.

A *partition* of a set S, requires U to be a *partition topology*. A partition topology, is a topology whose basis β is complete decomposition into non-empty, disjoint subsets $\{b_i|i \in I\}$, called *blocks*, where I is an index set used to name different blocks, where:

$$\begin{aligned} &\text{(i) } \forall i \in I : b_i \neq \emptyset, \\ &\text{(ii) } \bigcup_{i \in I} b_i = \beta, \text{ and} \\ &\text{(iii) } \forall i, j \in I, i \neq j : b_i \cap b_j = \emptyset, \end{aligned} \tag{3.5}$$

We can equivalently represent a partition as a total and surjective function:

$$f : \beta \rightarrow I \tag{3.6}$$

such that the above constraints are satisfied. However, a spatial partition cannot be defined simply as a set-theoretic partition of the plane, that is, as a partition of $\mathbb{R}^2$ or as a function $f : \mathbb{R}^2 \rightarrow I$, for two reasons: first, f cannot be assumed to be total in general, and second, f cannot be uniquely defined on the borders between adjacent subsets of $\mathbb{R}^2$.

3.5 Spatial Mapping

One of the goals of our PLR partition definition is that it associates labels with spatial features in partitions. Intuitively, this is similar to marking or coloring different portions of a partition. In general, arbitrary identifying values, called *labels*, that model thematic information of any complexity should be able to be assigned to different points, lines, or regions within a partition so that components of partitions can be identified by their label. Thus, the set A of labels used to mark point, line, and region features in a partition determines the type of the PLR partition. We make no assumptions as to the structure or contents of a specific label. For example, a PLR partition showing the countries of Canada, the US, and Mexico such that each country is labeled with its name can have a type $A = \{Canada, US, Mexico\}$. Any area in a PLR partition that is not specifically labeled in a partition is given the $\perp$ label. Thus, in the PLR partition of Canada, the US, and Mexico, the area shown that does not belong to those countries is labeled with $\perp$ and the type of the partition is $A = \{Canada, US, Mexico, \perp\}$. All points with the $\perp$ label are considered to be in the exterior of a partition.

We cannot simply map the plane to the elements of a type A because the boundaries between two features cannot be identified as belonging to one feature or the other. Instead, shared boundaries between two features belong to both features and are thus labeled with the labels of both features. Therefore, a PLR partition of type A is defined by a *spatial mapping* that maps points in the plane to elements of the power set of A. Points in the plane that belong solely to region features are mapped to an element $e \in 2^A$ that correspond to an element in A (i.e., $e = \{l\} | l \in A$). For example, the region $R2$ in Fig. 3.4 consists of all points mapped to the label $\{R2\}$.

Since the exterior of a PLR partition is a region that includes the unbounded face, points and lines always intersect at least one region or a boundary. For example, line $L1$ in Fig. 3.4 intersects region $R1$, the exterior of the partition, and the boundary of region $R1$. Therefore, each point that belongs to a line feature also belongs to at least one other component in the PLR partition, and must carry the labels of all components that contain it. This is also true for point features. Thus, boundaries between features are mapped to the labels of all features that surround the boundary, and intersecting features take the labels of all participating features. It follows that a PLR partition of type A is defined by a function $\pi : \mathbb{R}^2 \rightarrow 2^A$ that maps points to the power set of the set of labels that defines the type of the partition. For example, Fig. 3.4a depicts a scene where each feature is annotated with a name. Figure 3.4b

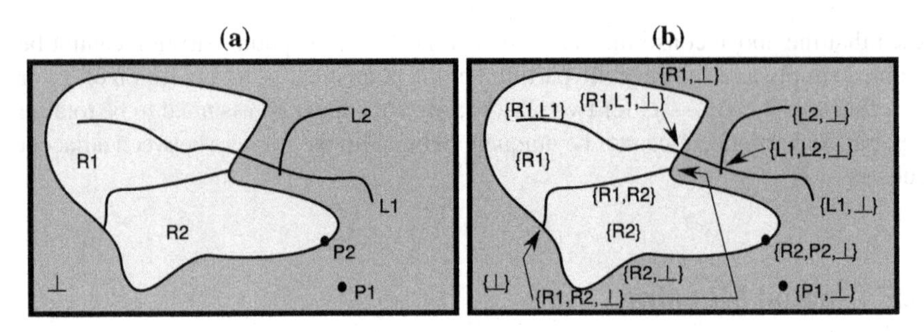

Fig. 3.4 A scene (**a**) with structured identified, and its labels according to a spatial mapping (**b**)

shows the result of the spatial mapping that corresponds to this scene in which each component is labeled.

As was mentioned previously, any points in a PLR partition that are not explicitly labeled are considered to be in the *exterior* of the partition and are labeled with the *undefined element* or *empty label* $\bot_A \in A$. If no ambiguities arise, we sometimes omit the type subscript and simply use $\bot$. We define a *spatial mapping* as the following:

Definition 3.1 A *spatial mapping* of type A is a total function $\pi : \mathbb{R}^2 \rightarrow 2^A$.

Therefore, a spatial mapping partitions the plane into blocks of identically labeled points. However, blocks defined in this way are not restrictive enough for our purposes. Spatial data types are typically defined not merely as point sets, but are in some sense regular; i.e., regions are not allowed to have cuts or punctures, etc. The topological concept of regular open sets [24] models this well. Therefore, we choose to model regions as two-dimensional, regular open point sets, and lines as one-dimensional, regular open point sets (points are trivial). This implies that a given spatial mapping may not necessarily represent a valid PLR partition. Therefore, we must define the properties a spatial mapping must have if it is supposed to represent a PLR partition.

3.6 PLR Partitions

In this section, we show how to identify each feature in a spatial mapping. We then provide constraints on these features in order to define the type of PLR partitions.

3.6.1 Components of PLR Partitions

Identifying features in a spatial mapping is somewhat non-intuitive. Consider region $R1$ in Fig. 3.4b and the portion of line $L1$ that intersects it. The points that make up $R1$ are associated with one of two blocks, the block labeled $\{R1\}$ and the block labeled $\{R1, L1\}$. To find the point set consisting of all points that define the region $R1$, we cannot simply use the function $\pi^{-1}(\{R1\})$, since it returns a point set containing a cut (the block labeled $\{R1, L1\}$ will not be returned). We make the observation that although points in $R1$ have one of two different labels, $\{R1\}$ is a subset of both labels. Therefore, we say that the label $\{R1\}$ is the *discriminant label* of the region, since it can be used to distinguish all points that make up that region. We now need a method to identify and retrieve blocks from a spatial mapping based on discriminant labels.

In order to identify the different features of a partition based on their discriminant labels, we take advantage of the property of regular open sets in the plane that every point that is a member of a regular open set is contained in a *neighborhood* of points that are also members of that set. We define the neighborhood of a given point $p = (p_x, p_y)$ in two dimensions as the set of points forming a circle around p with an infinitesimally small radius r. In order to retrieve the neighborhood of a point p, we define the neighborhood operation which returns the set of points contained in the neighborhood of p:

$$\begin{aligned} &N := \mathbb{R}^2 \rightarrow 2^{\mathbb{R}^2} \\ &N(p) = \{(x, y) | (x - p_x)^2 + (y - p_y)^2 = (r)^2\} \end{aligned} \tag{3.7}$$

We can now determine if a point in a partition belongs to a point, line, or region feature based on the properties of that point's neighborhood. For example, a point p that belongs solely to a region (as opposed to the intersection of a region and a line) will have the same label as every point in its neighborhood. In other words, the label of p will be equivalent to the union of the labels of all points in $N(p)$.

The predicate *isBasicRegion* takes a point and returns a value of *true* if it belongs solely to a region feature, and *false* otherwise. For instance, all points in the interior of region $R2$ in Fig. 3.4b are basic region points.

$$\begin{aligned} &isBasicRegion : \mathbb{R}^2 \rightarrow \mathbb{B} \\ &isBasicRegion(p) := \left(\pi(p) = \bigcup_{q \in N(p)} \pi(q) \right) \end{aligned} \tag{3.8}$$

In order to determine if a point p belongs to a region in general (and not a boundary between regions) regardless of how many line and point features also include that point, we must determine that all points in the neighborhood of p that happen to be identified as basic region points have the same label (i.e., the set containing the labels of all basic region points in $N(p)$ has a cardinality equal to 1). If p is on a boundary between regions, then its neighborhood will contain basic region points from each differently labeled region.

$$\begin{aligned} &isRegion : \mathbb{R}^2 \rightarrow \mathbb{B} \\ &isRegion(p) := |\{\pi(q) | q \in N(p) \wedge isBasicRegion(q)\}| = 1 \end{aligned} \tag{3.9}$$

For a spatial mapping π of type A, we also define the *discriminant spatial mapping* for regions $\pi_r : \mathbb{R}^2 \rightarrow 2^A$ that maps a point to its discriminant region label. For instance, a point p that lies on the intersection of region $R1$ and line $L1$ in Fig. 3.4a is a region point but not a basic region point, and $\pi_r(p) = \{R1\}$.

$$\begin{aligned} &\pi_r : \mathbb{R}^2 \to 2^A \\ &\pi_r(p) = \{\pi[q] | q \in N(p) \wedge \mathit{isBasicRegion}(q) \wedge \mathit{isRegion}(p)\} \end{aligned} \tag{3.10}$$

Note that the discriminant spatial mapping for regions will map points that lie on the boundary between two regions to the empty set. Thus, the blocks that contain labels in the discriminant spatial mapping for regions will be only the interiors of the regions in the spatial mapping.

A region boundary point is characterized as belonging to multiple regions; therefore, a point p lies on a region boundary if at least two points in its neighborhood are basic region points from different regions (i.e., the set of basic region labels belonging to points in $N(p)$ must contain at least two elements). We define the predicate *isRegionBound* to test if a point lies on a region boundary. For example, point p labeled $\{R2, P2, \perp\}$ in Fig. 3.4 is a region boundary because $N(p)$ contains basic region points from both $R2$ and the exterior region $\perp$

$$\begin{aligned} &\mathit{isRegionBound} : \mathbb{R}^2 \to \mathbb{B} \\ &\mathit{isRegionBound}(p) := |\{\pi(q) | q \in N(p) \wedge \mathit{isBasicRegion}(q)\}| > 1 \end{aligned} \tag{3.11}$$

For a spatial mapping π of type A, we also define the discriminant spatial mapping for region boundaries $\pi_{rb} : \mathbb{R}^2 \to 2^A$ that maps a point to its region boundary label. If the point is not on a region boundary, it is mapped to the empty set. For example, point p labeled $\{R2, P2, \perp\}$ in Fig. 3.4 lies on a region boundary and $\pi_{rb}(p) = \{R2, \perp\}$.

$$\begin{aligned} &\pi_{rb} : \mathbb{R}^2 \to 2^A \\ &\pi_{rb}(p) = \{\pi[q] | q \in N(p) \wedge \mathit{isBasicRegion}(q) \wedge \mathit{isRegionBound}(p)\} \end{aligned} \tag{3.12}$$

Now that we can discriminate the regions and region boundaries in a PLR partition regardless of the presence of point and line features, we turn our attention to identifying the point and line components themselves. Point features are unique in that the neighborhood of a point feature will not contain the discriminant label of the point. Therefore, to determine if point p is a point feature, we compute the difference of each label in $N(p)$ with the label of p. The resulting label will either be the discriminant label of p, if it is a point feature, or the empty set if p is not a point feature. For example, point $P1$ and point $P2$ in Fig. 3.4 are mapped to a label that contains an identifier that no other points in their neighborhood contain, namely $P1$ and $P2$, respectively.

$$\begin{aligned} &\mathit{isPoint} : \mathbb{R}^2 \to \mathbb{B} \\ &\mathit{isPoint}(p) := |\pi(p) - \bigcup_{q \in N(p)} \pi(q)| > 0 \end{aligned} \tag{3.13}$$

For a spatial mapping π of type A, we also define the discriminant spatial mapping for points $\pi_p : \mathbb{R}^2 \rightarrow 2^A$ that maps a point to the label that identifies it as a point feature. If a point is not a point component in the PLR partition, then the discriminant spatial mapping for points maps that point to the empty set.

$$\begin{aligned} &\pi_p : \mathbb{R}^2 \rightarrow 2^A \\ &\pi_p(p) = \pi(p) - \bigcup_{q \in N(p)} \pi(q) \end{aligned} \tag{3.14}$$

Lines introduce several difficulties when we attempt to identify line components of a PLR partition. Therefore, to begin identifying a line component, we must first remove the discriminant labels for points, regions, and region boundaries from a point or point set in question such that only identifiers in labels that pertain to line components remain; once the line components themselves are identifiable, we can then determine their structure. We define the label stripping function S that takes a point and returns the label of that point with any point, region, and region boundary discriminant labels removed:

$$\begin{aligned} &S := \mathbb{R}^2 \rightarrow 2^A \\ &S(p) = \pi(p) - \pi_r(p) - \pi_{rb}(p) - \pi_p(p) \end{aligned} \tag{3.15}$$

A pair of lines is able to be configured in many of the same ways that a pair of regions can be configured. For example, a pair of lines can be disjoint, overlap, etc. However, lines have the interesting property in PLR partitions that the boundary of a line is not necessarily shared with another line. A region's boundary is always shared with another region, sometimes the exterior region, in a PLR partition, but a line has no such constraint. Thus, line boundaries cannot be determined in the same way as region boundaries. Another problem is that if two lines cross each other at a point ($L1$ and $L2$ cross at a point in Fig. 3.4), the point at which they cross is actually an interior point of each line, not a boundary within each line; thus, such a point actually has two discriminant labels for the two lines whose interiors' it lies in. Furthermore, the inverse of the discriminant spatial mapping must return the point in question if given either of the discriminant labels for the intersecting lines. Effectively, a point lies on the interior of a line if it has at least two points in its neighborhood that share its discriminant label for the line to which it belongs:

$$\begin{aligned} &isLine : \mathbb{R}^2 \rightarrow \mathbb{B} \\ &isLine(p) := \exists q, s \in N(p) | q \neq s \wedge S(q) = S(s) \wedge S(q) \neq \emptyset \wedge \pi(q) \subseteq \pi(p) \end{aligned} \tag{3.16}$$

For a spatial mapping π of type A, we also define the discriminant spatial mapping for lines $\pi_l : \mathbb{R}^2 \rightarrow 2^A$ that maps a point to the label that identifies it as a line feature. If a point is not a line component in the PLR partition, then the discriminant spatial mapping for lines maps that point to the empty set. This mapping takes into account the problem of an interior line point carrying the discriminant labels of multiple lines that intersect at a point, as mentioned above. Thus, a point is mapped to a label containing elements that discriminate all line interiors at that point (note that a line boundary is defined by a line not passing *through* a point, but emanating from a point). Similarly, the inverse mapping of a discriminant line label a must return all points such that a is a subset of that point's discriminant line labels.

$$\begin{aligned}
&\pi_l : \mathbb{R}^2 \to 2^A \\
&\pi_l(p) = \{S[q] | q, s \in N(p) \wedge isLine(p) \wedge S(q) = S(s) \wedge q \neq s\} \\
&\pi_l^{-1} : 2^A \to 2^{\mathbb{R}^2} \\
&\pi_l^{-1}(Y) = \{x \in \mathbb{R}^2 | isLine(x) \wedge Y \subseteq \pi_l(x)\}
\end{aligned} \tag{3.17}$$

A point p is on the boundary of a line if the line extends in only one direction from p; therefore, if there exists a point q in the neighborhood of p such that $S(q) \subseteq S(p)$ and no other point in p's neighborhood contains the same stripped label as q, then point p is a boundary of a line. The same logic identifies the boundaries of two lines (or more) that meet at a single point.

$$\begin{aligned}
&isLineBoundary := \mathbb{R}^2 \to \mathbb{B} \\
&isLineBoundary(p) = \exists q \in N(p) | S(q) \subseteq S(p) \wedge S(q) \neq \emptyset \\
&\qquad \wedge (\nexists s \in N(p) | q \neq s \wedge S(q) = S(s))
\end{aligned} \tag{3.18}$$

For a spatial mapping π of type A, we also define the discriminant spatial mapping for line boundaries $\pi_{lb} : \mathbb{R}^2 \to 2^A$ that maps a point to the label that identifies it as a line boundary feature. If a point is not a line boundary component in the PLR partition, then the discriminant spatial mapping for line boundaries maps that point to the empty set. The inverse of the discriminant spatial mapping for line boundaries maps a discriminant line label to all boundary points of that line

$$\begin{aligned}
&\pi_{lb} : \mathbb{R}^2 \to 2^A \\
&\pi_{lb}(p) = \{S[q] | (\nexists s \in N(p) | q \neq s \wedge S(q) = S(s)) \\
&\qquad \wedge isLineBoundary(p) \wedge S(q) \neq \emptyset\} \\
&\pi_{lb}^{-1} := 2^A \to 2^{\mathbb{R}^2} \\
&\pi_{lb}^{-1}(Y) = \{x \in \mathbb{R}^2 | isLineBoundary(x) \wedge Y \subseteq \pi_{lb}(x)\}
\end{aligned} \tag{3.19}$$

3.6.2 PLR Partition Definition

The features in a spatial mapping are identified based on the predicates and the discriminant mappings defined above. Similarly to a spatial partition, we identify the blocks representing regions, region boundaries, lines, line boundaries, and points in spatial mapping, and then impose constraints on those blocks and their labels to define a PLR partition. Let π be a spatial mapping of type A. The features in π are:

$$\begin{array}{lll}
\text{(i)} & \rho(\pi) := \pi_r^{-1}(2^A) & (regions) \\
\text{(ii)} & \omega_\rho(\pi) := \pi_{rb}^{-1}(2^A) & (region\ borders) \\
\text{(iii)} & \lambda(\pi) := \pi_l^{-1}(2^A - \{\emptyset\}) & (lines) \\
\text{(iv)} & \omega_\lambda(\pi) := \pi_{lb}^{-1}(2^A - \{\emptyset\}) & (line\ borders) \\
\text{(v)} & \varphi(\pi) := \pi_p^{-1}(2^A - \{\emptyset\}) & (points) \\
\text{(vi)} & \pi_\rho^\circ := \bigcup_{r \in \rho(\pi) | \pi[r] \neq \{\bot\}} r & (interior\ of\ regions) \\
\text{(vii)} & \pi_\lambda^\circ := \bigcup_{l \in \lambda(\pi)} l & (interior\ of\ lines) \\
\text{(viii)} & \pi_\varphi^\circ = \emptyset & (interior\ of\ points) \\
\text{(ix)} & \partial \pi_\rho := \bigcup_{b \in \omega_\rho(\pi)} b & (region\ boundary) \\
\text{(x)} & \partial \pi_\lambda := \bigcup_{b \in \omega_\lambda(\pi)} b & (line\ boundary) \\
\text{(xi)} & \partial \pi_\varphi := \bigcup_{p \in \varphi(\pi)} p & (point\ boundary) \\
\text{(xii)} & \pi_\rho^- := \pi_r^{-1}(\{\bot\}) & (exterior\ of\ regions) \\
\text{(xiii)} & \pi_\lambda^- := \mathbb{R}^2 - \pi_\lambda^\circ - \partial \pi_\lambda & (exterior\ of\ lines) \\
\text{(xiv)} & \pi_\varphi^- := \mathbb{R}^2 - \partial \pi_\varphi & (exterior\ of\ points)
\end{array} \tag{3.20}$$

A *PLR partition* is defined as a spatial mapping π of type A such that:

1. The regions of π are two-dimensional regular open point sets.
2. The borders of regions have a discriminant label equal to the union of discriminant labels of all adjacent regions.
3. The lines of π are one-dimensional regular open point sets.
4. The borders of lines have a discriminant label equal to the union of discriminant labels of all adjacent lines.

More formally:

Definition 3.2 A *PLR partition* is a spatial mapping π of type A such that:

(i) $\forall r \in \rho(\pi) : r = \overline{r}^\circ$
(ii) $\forall b \in \omega_\rho : \pi_{rb}[b] = \{\pi_r[[r]] | r \in \rho(\pi) \wedge b \subseteq \partial r\}$
(iii) $\forall l \in \lambda(\pi) : l = \overline{l}^\circ$
(iv) $\forall b \in \omega_\lambda : \pi_{lb}[b] = \{\pi_l[[l]] | l \in \lambda(\pi) \wedge b \subseteq \partial l\}$.

3.7 Using PLR Partitions to Define Points, Lines, and Regions

Because PLR partitions contain structures that correspond to complex points, lines, and regions, they can be used to provide a concise definition of both simple and complex spatial types. Indeed, each case of a simple or complex spatial data type is simply a special case of a PLR partition. Regions, in particular, are a special case of both PLR partitions and spatial partitions. A point, line, or region will be defined by a spatial mapping π of type A in which A will contain exactly two labels. We will use the $\bullet$ symbol to indicate the label of the spatial data type being defined, and the

$\perp$ symbol to indicate the exterior region label; therefore, $A = \{\perp, \bullet\}$ for all spatial data types in the following definitions. In practice, any label can be used in place of $\bullet$.

A complex region is a special case of a PLR partition in which the partition contains no point or line features, and there are only two region labels, one of which is the exterior label. Thus, the PLR partition contains only one block corresponding to a single region.

Definition 3.3 A *complex region* is a PLR partition π of type A in which:

(i) $\lambda(\pi) = \emptyset$
(ii) $\varphi(\pi) = \emptyset$
(iii) $|\rho(\pi)| = 2$
(iv) $A = \{\perp, \bullet\}$

A simple region is then a complex region containing a single *face*. A face of a region is a connected component. A connected component of an open set S is a maximal subset $T \subseteq S$ such that any two points in T can be connected by a curve that lies completely within T. T is maximal if there are no other points in S that can be added to T such that T remains connected. An indexed set, denoted $\{\}_I$, is a set such that the elements of the set are ordered by some index. We will use indexed sets ordered by non-negative integers. Let r be an open point set in the plane. We define the set of maximal connected components in r to be the indexed set $\gamma(r)$:

$$\gamma(r) = \{c_0, \ldots, c_n | c_i \subseteq r \wedge c_i \textit{ is a maximal connected subset of } r\}_I \quad (3.21)$$

A simple region is a complex region whose interior is a single maximal connected point set.

Definition 3.4 A *simple region* is a PLR partition π of type A in which:

(i) $\lambda(\pi) = \emptyset$
(ii) $\varphi(\pi) = \emptyset$
(iii) $|\rho(\pi)| = 2$
(iv) $A = \{\perp, \bullet\}$
(v) $|\gamma(\pi_\rho^\circ))| = 1$

A complex line is a special case of a PLR partition such that the partition contains no region or point features, and there are only two labels in the set defining the type of the partition: the exterior region label and the label of the line. The only region in the partition is the exterior region:

Definition 3.5 A *complex line* is a PLR partition π of type A in which:

(i) $\rho(\pi) = \pi_\rho^-$
(ii) $\varphi(\pi) = \emptyset$
(iii) $A = \{\perp, \bullet\}$
(iv) $|\lambda(\pi)| = 1$

A simple line is a complex line that has only two boundary point. Having only two boundary points implies that the line's interior is a single maximally connected point set and there are no branches in the line:

Definition 3.6 A *simple line* is a PLR partition π of type A in which:

(i) $\rho(\pi) = \pi_{\rho}^{-}$
(ii) $\varphi(\pi) = \emptyset$
(iii) $A = \{\perp, \bullet\}$
(iv) $|\lambda(\pi)| = 1$
(v) $|\partial \pi_{\lambda}| = 2$

A complex point is a PLR partition containing only a single region feature, the exterior region, no point features, and a single set of point features that have the same the label. Thus, there are only two labels in the set defining the type of the partition that represents a complex point:

Definition 3.7 A *complex point* is a PLR partition π of type A in which:

(i) $\rho(\pi) = \pi_{\rho}^{-}$
(ii) $\lambda(\pi) = \emptyset$
(iii) $A = \{\perp, \bullet\}$
(iv) $|\partial \pi_{\varphi}| > 0$

A simple point is then a complex point that has only a single boundary point (recall point interiors are empty).

Definition 3.8 A *simple point* is a PLR partition π of type A in which:

(i) $\rho(\pi) = \pi_{\rho}^{-}$
(ii) $\lambda(\pi) = \emptyset$
(iii) $A = \{\perp, \bullet\}$
(iv) $|\partial \pi_{\varphi}| = 1$

The fact that the traditional types can be expressed in terms of spatial partitions and PLR partitions is significant in the sense that any operations defined over spatial partitions and PLR partitions can be applied to the traditional types as well. Furthermore, if we prove type closure over those operations, as we do in the following chapters, those operations also have type closure over the traditional spatial data types.

3.8 Summary

PLR partitions represent maps containing point, line, and region features; furthermore, PLR partitions allow us to define the traditional data types in a concise way. Because the traditional spatial types are special cases of PLR partitions, they inherit any operations defined for PLR partitions. Thus, PLR partitions provide an interesting lens through which we can study the traditional types and operations over them.

Although PLR partitions represent networks in the form of complex lines, the model lacks the ability to inherently model concepts familiar in road networks, such as one way streets, intersection with turn restrictions, and roads that cross but do not form an intersect (e.g., overpasses). It is possible to encode such information as attribute values, but an extension to the data type to fundamentally represent transportation network constructs would create the opportunity to define new classes of operations over the data type itself.

References

1. Broome, F.R., Meixler, D.B.: The TIGER Data Base Structure. Cartography and Geographic Information Systems **17**(1), 39–48 (1990)
2. Clementini, E., Di Felice, P.: A Model for Representing Topological Relationships between Complex Geometric Features in Spatial Databases. Information Systems **90**, 121–136 (1996)
3. Clementini, E., Felice, P.D.: Topological invariants for lines. Knowledge and Data Engineering, IEEE Transactions on **10**(1), 38–54 (1998)
4. Dugundi, J.: Topology. Allyn and Bacon (1966)
5. Egenhofer, M.J.: Definitions of line-line relations for geographic databases. IEEE Data Eng. Bull. **16**(3), 40–45 (1993)
6. Egenhofer, M.J., Herring, J.: Categorizing binary topological relations between regions, lines, and points in geographic databases. University of Maine, Orono, Maine, Dept. of Surveying Engineering, Technical Report (1990)
7. Erwig, M., Schneider, M.: Partition and conquer. In: Spatial Information Theory A Theoretical Basis for GIS, pp. 389–407. Springer (1997)
8. Filho, W.C., de Figueiredo, L.H., Gattass, M., Carvalho, P.C.: A Topological Data Structure for Hierarchical Planar Subdivisions. In: 4th SIAM Conference on Geometric Design (1995)
9. Fisher, H.: Symap. In: Projects: 1960–1970. Laboratory for Computer Graphics and Spatial Analysis (1966)
10. Güting, R.H.: Geo-Relational Algebra: A Model and Query Language for Geometric Database Systems. In: Extending Database Technology, pp. 506–527 (1988)
11. Güting, R.H., Schneider, M.: Realm-based spatial data types: the rose algebra. The VLDB JournalThe International Journal on Very Large Data Bases **4**(2), 243–286 (1995)
12. Herring, J.: TIGRIS: Topologically Integrated Geographic Information Systems. In: 8th International Symposium on Computer Assisted Cartography, pp. 282–291 (1987)
13. Herring, J.: TIGRIS: A Data Model for an Object-Oriented Geographic Information System. Computers and Geosciences **18**(4), 443–452 (1992)
14. Huang, Z., Svensson, P., Hauska, H.: Solving Spatial Analysis Problems with GeoSAL, A Spatial Query Language. In: Proceedings of the 6th Int. Working Conf. on Scientific and Statistical Database Management, pp. 1–17. Institut f. Wissenschaftliches Rechnen Eidgenoessische Technische Hochschule Zürich (1992)
15. Ledoux, H., Gold, C.: A voronoi-based map algebra. Progress in Spatial Data Handling, pp. 117–131 (2006)
16. McKenney, M., Schneider, M.: Advanced operations for maps in spatial databases. Progress in Spatial Data Handling, pp. 495–510 (2006)
17. McKenney, M., Schneider, M.: Plr partitions: a conceptual model of maps. In: Advances in Conceptual Modeling–Foundations and Applications, pp. 368–377. Springer (2007)

18. McKenney, M., Schneider, M.: Spatial Partition Graphs: A Graph Theoretic Model of Maps. In: Int. Symp. on Spatial and Temporal Databases (SSDT), pp. 167–184 (2007)
19. Mennis, J.: Multidimensional map algebra: Design and implementation of a spatio-temporal gis processing language. Transactions in GIS **14**(1), 1–21 (2010)
20. Schneider, M.: Spatial Data Types for Database Systems - Finite Resolution Geometry for Geographic Information Systems, vol. LNCS 1288. Springer-Verlag, Berlin Heidelberg (1997)
21. Schneider, M., Behr, T.: Topological relationships between complex spatial objects. ACM Trans. Database Syst. **31**(1), 39–81 (2006). DOI 10.1145/1132863.1132865.
22. Scholl, M., Voisard, A.: Thematic Map Modeling. In: SSD '90: Proceedings of the first symposium on Design and implementation of large spatial databases, pp. 167–190. Springer-Verlag New York, Inc., New York, NY, USA (1990)
23. Steinitz, C., Parker, P., Jordan, L.: Hand-drawn overlays: Their history and prospective uses. Landscape architecture **66**(5), 444–455 (1976)
24. Tilove, R.B.: Set membership classification: A unified approach to geometric intersection problems. Computers, IEEE Transactions on **100**(10), 874–883 (1980)
25. Tomlin, C.D.: A map algebra. Harvard Graduate School of Design (1990)
26. Tomlin, C.D.: Map algebra: one perspective. Landscape and Urban Planning **30**(1), 3–12 (1994)
27. Voisard, A., David, B.: Mapping Conceptual Geographic Models onto DBMS Data Models. Tech. Rep. TR-97-005, International Computer Science Institute (1997)
28. Worboys, M.F., Bofakos, P.: A Canonical Model for a Class of Areal Spatial Objects. In: 3rd Int. Symp. on Advances in Spatial Databases, pp. 36–52. Springer-Verlag (1993)

Chapter 4
Foundational Operations for Maps

A complete Map Framework requires both data types and operations over the data types. In this chapter, we introduce the three foundational operations over maps, known as the *power operations* for maps, that are closed under the type of spatial partitions. The power operations are significant because all map operations can be expressed in terms of the three power operations, and since they are closed under the type of spatial partitions, the output of a power operation is always a spatial partition. The power operations allow one to (i) combine two spatial partitions, (ii) distinguish all faces of all regions in a spatial partition, and (iii) alter label values. Thus the power operations allow users to create new partitions by combining existing partitions, identify the components (i.e., regions) in a partition, and alter the partition, since regions are defined based on their labels. In later chapters, we will use the power operations to define map operations found in the literature. The power operations were originally defined in [1] and used extensively in [2, 3]. The power operations are named *intersection*, *relabel*, and *refine* and are explained in the remainder of this chapter.

4.1 Intersection

The intersection operation allows a user to construct a new spatial partition by combining two existing spatial partitions. The resulting spatial partition is *area inclusive* of the existing partitions; meaning that any area covered by a region in the input spatial partitions is covered by a region in the result spatial partition. This differs from traditional conceptions of the geometric set intersection, in which the result set of the intersection of two regions defined as point sets only contains elements contained in both input sets. Furthermore, a point in the intersection of two spatial partitions will have the labels of the equivalent point in both input partitions. The labels of the result partition will be constructed such that the original partitions may be extracted from the result of the intersection; therefore, the intersection of two spatial partitions does not loose any information.

M. McKenney and M. Schneider, *Map Framework*,
DOI 10.1007/978-3-319-46766-5_4

4.1.1 Constructing the Intersection Operation

Let π be a spatial partition of type A and σ be a spatial partition of type B. The intersection of π and σ must be constructed as a spatial mapping that maps points in the plane to a combination of labels from A and B. Thus, the intersection $\pi \cap \sigma$ will be of type $A \times B$. Therefore, the labels of $\pi \cap \sigma$ will be sets of *tuples*, specifically pairs, such that the first element in each tuple will be a label from A and the second element in each tuple will be a label from B. Recall that in a spatial partition, a region label is a set containing one element; this carries into the intersection where a region in the intersection of two spatial partitions will contain a single tuple. Boundaries between regions will contain the labels (each label will be a tuple under intersection) of all regions.

Because a spatial partition is defined by a spatial mapping, we can define the intersection $\pi \cap \sigma$ by a spatial mapping based on the spatial mappings of π and σ. Spatial mappings are total, so every point in the plane is mapped to a label by both π and σ. The spatial mapping for the intersection of π and σ then must map each point in the plane to the label defined as a set of pairs such that the first element in each pair is a label in A and the second element in each pair is a label in B. Therefore, the intersection of two spatial partitions $\pi \cap \sigma$ is a spatial mapping $\theta : \mathbb{R}^2 \to 2^{A \times B}$ that satisfies the constraints of a spatial partition. For a spatial partition of type A, let $[A]$ denote the set of all possible spatial partitions of type A; in other words, $[A]$ defines the data type of spatial partitions of type A.

The spatial mapping representing the intersection $\pi \cap \sigma$ is constructed by mapping each point p in the plane to a subset of the cartesian product of the labels that p maps to under both π and σ. Figure 4.1 depicts two spatial partitions and their intersection. Let π be the spatial partition of type A in Fig. 4.1a, let σ be the spatial partition of type B in Fig. 4.1b, and let θ be the intersection $\pi \cap \sigma$ of type $A \times B$ shown in Fig. 4.1c. $p1, p2, p3, p4$, and $p5$ indicate interesting points in the partitions that we will use to illustrate the intersection operation.

Constructing the spatial mapping for $p1$ in θ is relatively straightforward since it lies within a region in both π and σ. $\pi(p1) = \{\bot_A\}$ and $\sigma(p1) = \{x\}$. The cartesian product $\pi(p1) \times \sigma(p1) = \{(\bot_A, x)\}$. In Fig. 4.1c, $p1$ clearly lies in the interior of the exterior region of π and region $\{x\}$ of σ, thus, $p1$ lies in a region that is equivalent to the intersection of those complex regions, and the label contains the information from labels of all regions in π and σ that contain $p1$. No information is lost from the original partitions, and the label for $p1$ in θ contains a single element. Constructing the labels for points in the intersection of two spatial partitions that do not lie on a boundary of either input spatial partition always behaves in such a manner.

A boundary point in a spatial partition contains the labels of all adjacent regions, and the label is a set with cardinality greater than 1. Constructing the labels for points that lie on a boundary in a spatial partition is not quite so straightforward as points that do not lie on boundaries because the cartesian product of the labels of a point that lies on the boundary of one input region to an intersection operation may produce tuples indicating a region borders that point that does not actually border the point in

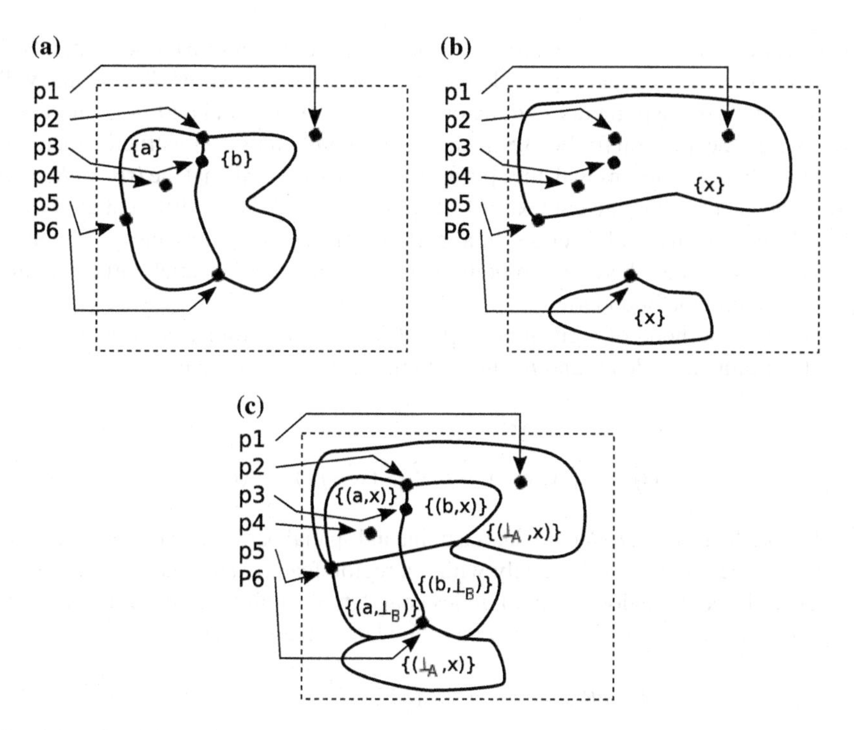

Fig. 4.1 **a** Spatial partition π of type A consists of two regions. **b** Spatial partition σ of type B contains a single region with two faces. **c** The intersection of partition A and B results in spatial partition θ of type $A \times B$. In all three figures, five points of interest are labeled

the intersection of the spatial partitions. For example $p6$ in Fig. 4.1 indicates a point that lies on the boundary of regions $\{a\}$, $\{b\}$, and $\{\bot_A\}$ in π and regions $\{x\}$ and $\{\bot_B\}$ in σ. Thus:

$$\begin{aligned} \pi(p6) &= \{a, b, \bot_A\} \\ \sigma(p6) &= \{x, \bot_B\} \\ \pi(p6) \times \sigma(p6) &= \{(a, x), (a, \bot_B), (b, x), (b\bot_B), (\bot_A, x), (\bot_A, \bot_B)\} \end{aligned} \tag{4.1}$$

However, $p6$ does not actually border a region with label $(\bot_A, \bot_B)$ in Fig. 4.1c. Therefore, the cartesian product of the labels of a point that lies on the boundary of one of the input regions to an intersection operation indicates the possible regions the labels of regions that are adjacent to that point in the intersection, but not all of those regions may exist or bound that point in the intersection. To discover which regions actually bound a point that lies on a boundary in the result of an intersection operation, we use concepts developed in Chap. 3 to identify if a point in a spatial partition lies on the interior of a region. Essentially, we can determine the regions that will bound any point p in the intersection of two spatial partitions π and σ by examining the labels of points that lie in the *neighborhood* of p but do not lie on a boundary of either π or σ. Equation 3.7 defines a function to return the set of points in the neighborhood of p and Eq. 3.8 defines a predicate named *isBasicRegion* that will evaluate to *True* if a point does not lie on a boundary of a spatial partition, and *False* otherwise. The notation for Eq. 3.8 must change slightly for our construction here, so we repeat both definitions with the appropriate changes. Recall that the *isBasicRegion* predicate is defined in the context of PLR partitions, and thus is differentiated from the *isRegion* predicate. Under a spatial partition all regions satisfy the definition of a basic regions; i.e., there are no point and line features in a spatial partition, only regions and region boundaries.

The neighborhood of a given point $p = (p_x, p_y)$ in two dimensions as the set of points forming a circle around p with an infinitesimally small radius r.

$$\begin{aligned} &N := \mathbb{R}^2 \rightarrow 2^{\mathbb{R}^2} \\ &N(p) = \{(x, y) | (x - p_x)^2 + (y - p_y)^2 = (r)^2\} \end{aligned} \tag{4.2}$$

The predicate *isBasicRegion* takes a point and spatial partition and returns a value of *True* if the point lies completely within a region feature, and *False* if it lies on a boundary between regions. A point p lies completely within a region if the labels of all points in the neighborhood $N(p)$ have an identical label to p.

$$\begin{aligned} &isBasicRegion := \mathbb{R}^2 \times [A] \rightarrow \mathbb{B} \\ &isBasicRegion(p, \pi) = \left(\pi(p) = \bigcup_{q \in N(p)} \pi(q)\right) \end{aligned} \tag{4.3}$$

The label for any point p in spatial partition $\theta = \pi \cap \sigma$ can then be constructed from the labels of points in the neighborhood of p that are basic regions:

$$\begin{aligned} &intersection : [A] \times [B] \rightarrow [A \times B] \\ &intersection(\pi, \sigma) = \theta \\ &\quad where\ \theta(p) = \{(\pi[q], \sigma[q]) | q \in N(p) \\ &\qquad\qquad \wedge\ isBasicRegion(q, \pi) \wedge\ isBasicRegion(q, \sigma)\} \end{aligned} \tag{4.4}$$

Figure 4.1 illustrates the intersection operation.

Using the intersection operation, we can now show how the labels for the points of interest identified in Fig. 4.1 are constructed. In π, the label of all points in the neighborhood of $p1$ is $\{\perp_A\}$ and the label of all points in the neighborhood of $p1$ in σ is $\{x\}$. $p1$ lies in the interior of a region in both input partitions, so its label will be a singleton set in the intersection with value $\{(\perp_A, x)\}$.

$p2$ lies on the boundary of three regions in π and the interior of a region in σ. In π, the points in the neighborhood of $p2$ that lie in region interiors are $\{a\}$, $\{b\}$, and $\{\perp_A\}$. In σ, $p2$ lies in the interior of a region, so all points in its neighborhood will have label $\{x\}$. When we collect the labels of points in the neighborhood of $p2$ from

Table 4.1 The labels for the points indicated in all three partitions in Fig. 4.1

	π	σ	θ
p1	$\{\bot_A\}$	$\{x\}$	$\{(\bot_A, x)\}$
p2	$\{a, b, \bot_A\}$	$\{x\}$	$\{(a, x), (b, x), (\bot_A, x)\}$
p3	$\{a, b\}$	$\{x\}$	$\{(a, x), (b, x)\}$
p4	$\{a\}$	$\{x\}$	$\{(a, x)\}$
p5	$\{a, \bot_A\}$	$\{x, \bot_B\}$	$\{(a, x), (a, \bot_B), (\bot_A, x), \bot_{A\times B}\}$
p6	$\{a, b, \bot_A\}$	$\{x, \bot_B\}$	$\{(a, \bot_B), (b, \bot_B), (\bot_A, x)\}$

both input spatial partitions, as in Eq. 4.4, we construct the tuples (a, x), (b, x), and $(\bot_A, x)$. Therefore, $p2$ lies on the boundary of three different regions in θ and its label will have cardinality 3 and will contain each of those tuples: $\{(a, x), (b, x), (\bot_A, x)\}$.

$p3$ through $p5$ are constructed in similar fashion. Recall that $p6$ created a situation that we must handle with care. Again, for every point in the neighborhood of $p6$ that is in the interior of a region in both π and θ, we construct a tuple containing those labels. Those labels indicate the regions whose boundaries contain $p6$ in θ. In this case, those tuples are: $(a, \bot_B)$, $(b, \bot_B)$, and $(\bot_A, x)$. Three regions meet at point $p6$ in θ; therefore, the cardinality of the label of $p6$ in θ is 3 and the label of $p6$ is: $\{(a, \bot_B), (b, \bot_B), (\bot_A, x)\}$. Table 4.1 contains the labels for all points of interest in all three partitions. In some cases, it is convenient to use the shorthand $\bot_{A\times B} = (\bot_A, \bot_B)$.

4.1.2 Type Closure

We must show that the intersection operation is closed under the type of spatial partitions. In other words, we must prove that if an intersection operation is computed over two spatial partitions, the result will be a spatial partition.

Lemma 4.1 *The intersection operation is closed under the type of spatial partitions.*

Proof The intersection operation, defined in Eq. 4.4, is constructed by creating a spatial mapping based upon two spatial mappings that meet the criteria of spatial partitions. Recall that a spatial partition is a spatial mapping where regions are regular open point sets and the boundaries contain the labels of all adjacent regions (Definition 2.2). When combining the spatial mappings of the input partitions, three cases arise: (i) a point lies in the interior of regions in both input spatial partitions, (ii) a point lies on the interior of a region from one spatial partition and on a boundary in the other spatial partition, and (iii) a point lies on a boundary in both spatial partitions. In case (i), it is clear that the point will lie on the interior of a region in the result spatial mapping, and that its label will contain a pair consisting of the regions from both input spatial partitions. In case (ii), the point p will lie on a boundary in the result partition, and thus, must have a label containing pairs indicating all regions that lie adjacent to that point; this label is computed by examining all points in the neighborhood of p that are contained in region interiors of the input partitions. Therefore, the point will contain the labels of all regions that bound p. Case (iii) follows the same logic as case (ii). Therefore, boundary points will always contain the labels of adjacent regions under intersection. Finally, we must ensure that the regions are regular open point sets. Because the mapping for the intersection operation is based on the neighborhood of points, and the input to the operations are spatial partitions, it follows from the definition of regular open point sets that we cannot introduce punctures or cuts into a region. In such a case, a point would have a different label from the points in the neighborhood. The only time this occurs is if a point is on a boundary, which has already been handled. Therefore, the intersection operation is closed under the type of spatial partitions. □

4.1.3 Alternate Construction of Intersection

Reference [1] presents an alternate construction of the intersection operation. We provide a summary for completeness.

The intersection of two partitions π and σ, of types A and B respectively, returns a spatial partition of type $A \times B$ such that each interior point p of the resulting partition is mapped to the pair $(\pi[p], \sigma[p])$, and all border points are mapped to the set of labels of all adjacent regions. Formally, the definition of intersection of two partitions π and σ of types A and B can be described in several steps. First, the regions of the resulting partition must be known. This can be calculated by a simple set intersection of all regions in both partitions, since $\cap$ is closed on regular open sets.

$$\rho_{\cap}(\pi, \sigma) := \{r \cap s | r \in \rho(\pi) \wedge s \in \rho(\sigma)\} \tag{4.5}$$

The union of all these regions gives the interior of the resulting partition:

$$\iota_{\cap}(\pi, \sigma) := \cup_{r \in \rho_{\cap}(\pi,\sigma)} r. \tag{4.6}$$

Next, the spatial mapping restricted just to the interior is calculated by mapping each interior point $p \in I = \iota_{\cap}(\pi, \sigma)$ to the pair of labels given by π and σ:

$$\pi_I := \lambda p : I.\{(\pi[p], \sigma[p])\} \tag{4.7}$$

Finally, the boundary labels are derived from the labels of all adjacent regions. Let $R := \rho_{\cap}(\pi, \sigma)$, $I := \iota_{\cap}(\pi, \sigma)$, and $F := \mathbb{R}^2 - I$. Then we have:

$$\begin{aligned} intersection &: [A] \times [B] \rightarrow [A \times B] \\ intersection(\pi, \sigma) &:= \pi_I \cup \lambda p : F.\{\pi_I[[r]] | r \in R \wedge p \in \overline{r}\} \end{aligned} \tag{4.8}$$

4.2 Relabel

The *relabel* operation provides a user with the ability to alter the labels in a spatial partition. For example, if labels are names of countries, and one country changes its name, the relabel operation allows a user to change the name in the spatial partition showing the countries. Although relabel is primarily oriented towards altering labels and attribute values, a relabel operation can change the structure of a spatial partition because regions are defined based on the labels. Specifically, if two regions are adjacent and share a boundary along a line, one can relabel those regions to the same label, and merge them into a single region.

4.2.1 Constructing the Relabel Operation

To relabel a partition π of type A, one applies a function $f : A \rightarrow B$ that alters the labels of π and effectively changes the type of the spatial partition. The function is user defined. The relabel operation is strictly limited to altering *region* labels, as opposed to boundary labels; this is key in order to enforce type closure.

In many cases, it is convenient to use an anonymous function in order to express a function used as a parameter for an operation; this situation is particularly relevant for the relabel operation. We use *lambda notation* to represent anonymous functions in the following form: $\lambda x : S.E(x)$ such that E is an expression that uses x. Lambda notation, in this form, is equivalent to the set expression $\{(x, E(x)) | x \in S\}$, which is a function. If the expression does not modify x, then it is assumed the expression returns x unchanged.

Formally, the relabel operation is defined as:

$$\begin{aligned} &relabel : [A] \times (A \rightarrow B) \rightarrow [B] \\ &relabel(\pi, f) := \lambda p : \mathbb{R}^2.\{f(x) | x \in \pi(p)\} \end{aligned} \tag{4.9}$$

The relabel operation has a structure that we denote the *replace if in* structure. In other words, for any label l, a relabel operation will pass each element $x \in l$ to the user defined function that will either leave x unchanged, or will replace x with an element of the range of the user defined function. It is important to note that from implementation purposes, a label may simply be an identifier to a record that contains actual attribute information.

In order to clarify terminology, we use the term *relabel* to refer to the relabel operation. A user defined function passed to a relabel operation is denoted a *relabeling function*.

The relabel operation is used to alter the thematic values of maps, as represented by labels, or to change the structure of maps through region *fusion*. For example, a spatial partition may contain two regions that share a boundary along a line; if a relabel operation alters the labels of those regions such that they are identical,

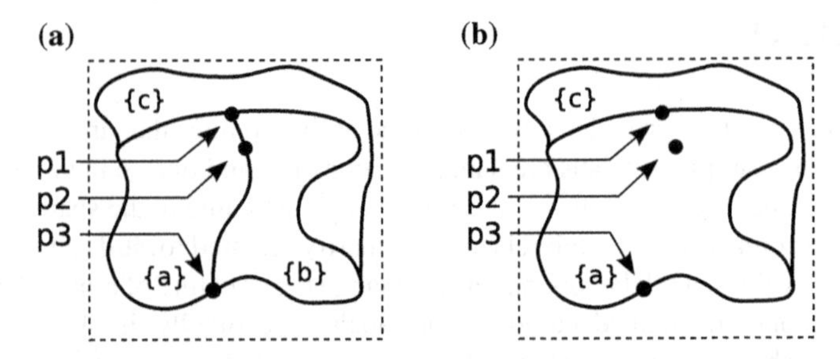

Fig. 4.2 **a** A spatial partition with points of interest marked. **b** The spatial partition after applying a relabel operation using the function in Eq. 4.10

Table 4.2 The labels of points of interest marked in Fig. 4.2a, b

	Labels in Fig. 4.2a	Labels in Fig. 4.2b
p1	$\{a, b, c\}$	$\{a, c\}$
p2	$\{a, b\}$	$\{a\}$
p3	$\{a, b, \perp\}$	$\{a, \perp\}$

the boundary between the regions disappears (more precisely, it becomes part of the interior of the fused region) and the two regions are fused together into a single larger region. Figure 4.2 depicts an example in which the relabeling function in Eq. 4.10 is applied to produce a new spatial partition. Equation 4.10 replaces the element b, in any label that contains b, with the element a. The effect of this function is that the labels of the region that were $\{b\}$ now become $\{a\}$, and the boundary labels for the points that lie on the border of the two regions change from $\{a, b\}$ to $\{a\}$; thus the boundary that previously separated region $\{a\}$ and region $\{b\}$ is absorbed into the interior of larger region $\{a\}$ in the result partition. The labels for the points of interest in Fig. 4.2 are shown in Table 4.2.

$$f(x) = \begin{cases} a & \text{if } x = b \\ x & \text{otherwise} \end{cases} \tag{4.10}$$

4.2.2 Type Closure

We must show that the relabel operation is closed under the type of spatial partitions. In other words, we must prove that if a relabel operation is computed over a spatial partition, the result will be a spatial partition.

Lemma 4.2 *The relabel operation is closed under the type of spatial partitions.*

Proof A spatial partition is defined as a spatial mapping whose regions are regular open point sets and whose boundaries have the labels of all adjacent regions

(Definition 2.2). We must show that the relabel operation applied to a spatial partition results in a spatial mapping that satisfies those constraints. The first constraint is rather straightforward to show. The relabel function operates only on labels, it does not directly take into account the points that map to those labels. Therefore, if a region in an input spatial partition is a regular open point set, and the label for that region is altered due to a relabel operation, then every point in the region will have an identically altered label; therefore, the point set will remain identical, and no cuts or punctures are introduced to it. It follows that every identically labeled point will be altered in the same way by the relabeling function because the relabeling function is a function by definition.

Finally, we must ensuring that boundary labels carry the labels of all adjacent regions. The *replace if in* structure of the relabeling function ensures that if a region labeled with element x has its label changed, then the boundary points of that region will also contain element x and will have element x changed in identical way to the corresponding region by the relabeling function. Furthermore, if the new element is identical to any other element in the boundary label, the cardinality of the boundary label set is effectively reduced by one, eliminating a boundary between two identically labeled regions. Therefore, the boundary will have the labels of all adjacent regions under a relabel operation. Thus, relabel is closed under the type of spatial partitions. □

4.3 Refine

A complex region may have multiple disconnected components, called faces [4]. One problem that occurs in spatial partitions is that there is not an inherent mechanism to identify the faces of a region separately from the entire region; the *refine* operation provides such functionality. Specifically, the refine operation will alter labels in a spatial partition such that each face in a region receives its own unique label; therefore, a spatial partition will only contain regions consisting of a single face after a user applies the refine operation.

4.3.1 Constructing the Refine Operation

The refine operation is constructed in a similar manner as the intersection operation. In essence, we will provide a transformation to a spatial mapping that defines a spatial partition that alters the labels that points map to. The label for a point will be altered by examining the region labels that lie in the neighborhood of that point. In this way, we can construct the refine operation with a single transformation to a spatial mapping.

In order to construct the refine operation, we need to identify the faces of regions that contain more than one face. Once the faces of a region are known, we will alter the label of each face by appending an identifier to it. In this book, we use integers as identifiers, but there is no implicit restriction on the type of identifiers.

A face of a region is a connected component. A connected component of an open set S is a subset $T \subseteq S$ such that any two points in T can be connected by a curve that lies completely within T. T is maximal if there are no other points in S that can be added to T such that T remains connected. An indexed set, denoted $\{\}_I$, is a set such that the elements of the set are ordered by some index. We will use indexed sets ordered by non-negative integers. Let r be a region defined as an open point set in the plane. We define the set of maximal connected components in r to be the indexed set $\gamma(r)$:

$$\gamma(r) = \{c_0, \ldots, c_n | c_i \subseteq r \wedge c_i \textit{ is a maximal connected subset of } r\}_I \tag{4.11}$$

Let π be a spatial partition of type A. The set of regions of π, denoted $\rho(\pi)$, are regular open point sets; therefore, any indexed set $\gamma(r)|r \in \rho(\pi)$ contains elements that are regular open point sets defining the faces of the region r. In order to refine a partition, we need to identify the regions in a partition that have more than one face, and we will need to identify each face in those regions. Let $\psi(\pi)$ be the set of all indexed sets containing connected components in the regions of π:

$$\psi(\pi) = \{\gamma(r)|r \in \rho(\pi)\} \tag{4.12}$$

We can now define the refine operation as a transformation on a spatial mapping. The refine operation is constructed in a similar fashion to the intersection operation. Each point in a spatial mapping is examined is altered to reflect the index of the face of the region it lies within, or, in the case of boundary points, to reflect the refined labels of adjacent faces.

We first examine the case in which a point lies in the interior of a face of a region. Figure 4.3 depicts a spatial partition and its refinement. $p3$, marked in the figure, is a point that lies completely within one face of the region $\{a\}$. Let r be the region containing $p3$. $p3$'s neighborhood contains all points within the same face of r; therefore, those points will all lie in the same element of $\gamma(r)$ (i.e., the same maximal connected subset of r, or the same face of r). Thus, we identify $p3$ as being part of a region face that must be refined, and we alter the label of $p3$ to a set containing a single pair where the first element in the pair is the original label element of $p3$, and the second element is the index of the point set containing $p3$ in $\gamma(r)$.

Points that lie on region boundaries must be handled more carefully, but again, we construct such labels based on their neighborhoods. A label of a boundary point must contain only the labels of all regions that bound that point. Therefore, we construct the label of a boundary point as set containing the label elements of all regions in the neighborhood of the point. This process happens to be identical to the process of refining a point that lies within a region face. Let π be a partition of type A. Formally:

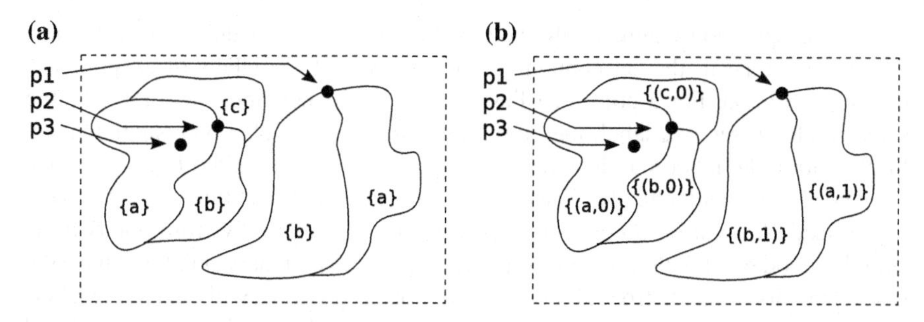

Fig. 4.3 **a** A spatial partition. **b** The spatial partition in part (**a**) after the refine operation is applied. Three points of interest are marked in each part of the figure

Table 4.3 The labels for points of interest marked in Fig. 4.3

	Labels in Fig. 4.3a	Labels in Fig. 4.3b
p1	$\{a, b, \bot\}$	$\{(a, 1), (b, 1), (\bot, 0)\}$
p2	$\{a, b, c\}$	$\{(a, 0), (b, 0), (c, 0)\}$
p3	$\{a\}$	$\{(a, 0)\}$

$$\begin{aligned}
&\mathit{refine} : [A] \rightarrow [A \times \mathbb{N}] \\
&\mathit{refine}(\pi) := \theta \textit{ where} \\
&\quad \theta(p) = \{ (\pi[q], i) | q \in N(p) \wedge \mathit{isBasicRegion}(p, \pi) \wedge q \in c_i \in X \in \psi(\pi) \}
\end{aligned} \tag{4.13}$$

For example, $p1$ in Fig. 4.3 lies on the boundary of a face of region $\{a\}$, a face of region $\{b\}$, and the exterior region (which contains only a single face in this instance). Therefore, according to the refine operation, it will have a label of $\{(a, 1), (b, 1), (\bot, 0)\}$, since those are the labels of its surrounding faces, and the refine operation constructs the labels of each point in the partition from the labels of its surrounding faces. Table 4.3 shows the labels of all points of interest in Fig. 4.3.

4.3.2 *Type Closure*

We must show that the refine operation is closed under the type of spatial partitions. In other words, we must prove that if a refine operation is computed over a spatial partition, the result will be a spatial partition.

Lemma 4.3 *The refine operation is closed under the type of spatial partitions.*

Proof The proof is similar to that of the intersect operation. A spatial partition is a spatial mapping with the restrictions that regions are regular open sets and boundaries contain the labels of all adjacent regions. The first criteria is straightforward. Let π be the input to the refine operator. Since the π is a spatial partition and its regions are regular open sets, we must simply not introduce cuts or punctures into regions.

The refine operation examines the neighborhood of a point, and assigns the new label to be equal to the set containing the elements of the labels of all points in the neighborhood. For a point that lies in a region, its neighbors will all have the same singleton set as a label, therefore, the new label will also be a singleton set. Furthermore, the label may be altered, but all points in its neighborhood will have their label altered identically. Thus, we cannot introduce punctures or cuts into a region face. A point on a boundary will construct its label only from points in its neighborhood that lie within regions. By definition, the boundary point will have the label elements of labels from all bounding regions. Again, we will alter some labels based on the refinement, but as with the first constraint, all points in the neighborhood of a region point will have their labels altered identically by refine. Therefore, the result of a refinement is a spatial partition. □

4.3.3 Alternate Construction of Refine

References [1, 2] presents an alternate construction of the refine operation. We provide a summary for completeness.

The *refine* operation is be defined in several steps. The regions of the spatial partition that results from the refine operation are the connected components of all regions of the original partition:

$$p_\gamma(\pi) := \bigcup_{r \in \rho(\pi)} \gamma(r)$$

The union of all these regions results in the interior of the resulting partition: $\iota_\gamma(\pi) := \bigcup_{r \in \rho_\gamma(\pi)} r$. This means that the set of interior and boundary points are not changed by refine.

We can now define the resulting partition on the interior:

$$\pi_I := \{(p, \{\pi[p], i)\}) | r \in \rho(\pi) \wedge \gamma(r) = \{c_1, \ldots, c_{n_r}\} \wedge i \in \{1, \ldots, n_r\} \wedge p \in c_i\}$$

Finally, we derive the labels for the boundary from the interior, much like the definition for intersection. Let $R := \rho_\gamma(\pi), I := \iota_\gamma(\pi)$, and $F := \mathbb{R}^2 - I$. Then:

$$\begin{aligned} \mathit{refine} &: [A] \rightarrow [A \times \mathbb{N}] \\ \mathit{refine}(\pi) &:= \pi_I \cup \lambda p : F.\{\pi_I[[r]] | r \in R \wedge p \in \overline{r}\} \end{aligned}$$

4.4 Closure of the Fundamental Map Operations

Finally, we state that the type of spatial partition is closed under the fundamental map operations. Corollary to Lemma 4.1 through Lemma 4.3:

Theorem 4.1 *The type of spatial partitions is closed under intersection, refine, and relabel.*

The closure of the fundamental map operations under the type of spatial partitions is significant because any operations constructed using these operations will maintain the closure properties of the underlying operations.

References

1. Erwig, M., Schneider, M.: Partition and conquer. In: Spatial Information Theory A Theoretical Basis for GIS, pp. 389–407. Springer (1997)
2. Erwig, M., Schneider, M.: Formalization of advanced map operations. In: 9th Int. Symp. on Spatial Data Handling, vol. 8, pp. 3–17 (2000)
3. McKenney, M., Schneider, M.: Advanced operations for maps in spatial databases. Progress in Spatial Data Handling, pp. 495–510 (2006)
4. Schneider, M., Behr, T.: Topological relationships between complex spatial objects. ACM Transactions on Database Systems (TODS) **31**(1), 39–81 (2006)

Chapter 5
Constructing Map Operations Using the Fundamental Map Operations

In Chap. 4, we developed the fundamental map operations, known as the *power operations* over maps, and showed that these operations are closed under the type of spatial partitions. In this chapter, we use the power operations to build additional map operations; because the new operations are constructed as compositions of the power operations, the new operations are also closed under the type of spatial partitions. We begin by briefly summarizing discussions of various operations in the literature. We then proceed to identify particular operations, explain them, and define them using the power operations. The composition of power operations into additional map operations has been also been explored in [4, 5, 11].

5.1 Identification of Operations

Many map operations are identified in the literature. In this section, we briefly indicate the most common operations. Explanations of the operations and their definition in terms of the Map Framework are provided in later sections. The most common operation in the literature is the *map overlay* [1, 3, 6–10, 12–15]. The overlay operation combines two maps into a single map; there are various versions of the operations will keep or discard portions of the input maps. The *reclassify* operation [1, 3, 9, 15] provides mechanisms to alter the thematic values of a map; this is largely achieved through the *relabel* operation in Map Framework. The *fusion* operation [2, 8–10, 12, 13] causes regions within a map to be fused together into larger regions. The *cover* operation [13] produces a single region indicating the area covered by a map. The *clipping* operation [13] allows a user to extract a specific area from a map as indicated by a query window. One can compute the *difference* [9] of two maps under multiple variations. The *superimpose* operation [2, 13] allows a user to combine two maps such that the area covered by one map *hides* any areas in the second map that are covered by the first. Finally, a the *window* operation [13] allows a user to extract regions from a map that are non-disjoint with some user-defined window.

M. McKenney and M. Schneider, *Map Framework*,
DOI 10.1007/978-3-319-46766-5_5

5.2 Overlay Operations

The overlay operation takes two spatial partitions and lays them over each other such that structures from both input partitions are visible in the result partition. There are two versions of this operation, which we respectively denote $overlay_1$ and $overlay_2$.

The $overlay_1$ operation produces a result in which only areas covered by both input partitions remain in the result; this is similar to a geometric intersection operation and, in fact, is defined using the intersection operation for spatial partitions. Figure 5.1 depicts two input partitions and the result of the $overlay_1$ operation.

Recall that under the intersection operation for spatial partitions, any area covered by a region in either input partition is covered by a region in the intersection of those partitions; therefore, we must relabel the intersection of two spatial partitions to achieve the $overlay_1$ operation. In order to remove any area in the intersection of two spatial partitions that was covered by a region in only a single input partition, we must relabel any region whose label contains a $\perp$ label to the exterior label of the intersection. For the intersection of two partitions respectively of type A and B, the exterior label will be $(\perp_A, \perp_B)$, or equivalently $\perp_{A\times B}$:

$$\begin{aligned}
&overlay_1 : [A] \times [B] \rightarrow [A \times B] \\
&overlay_1(\pi, \sigma) := relabel(\ intersection(\pi, \sigma), \\
&\qquad\qquad\qquad\qquad \lambda(a, b) : A \times B. \\
&\qquad\qquad\qquad\qquad\quad if\ a = \perp_A \vee b = \perp_b\ then\ \perp_{A\times B})
\end{aligned} \tag{5.1}$$

The second version of the overlay is identical to the intersection operation for spatial partitions. Under the $overlay_2$ operation, any area covered by a region in either input partition will be covered by a region in the output partition. Any area covered by a region from both input partitions will be covered by a region in the result with labels indicating both covering regions. Figure 5.2 depicts the result of an $overlay_2$ operation.

$$\begin{aligned}
&overlay_2 : [A] \times [B] \rightarrow [A \times B] \\
&overlay_2(\pi, \sigma) := intersection(\pi, \sigma)
\end{aligned} \tag{5.2}$$

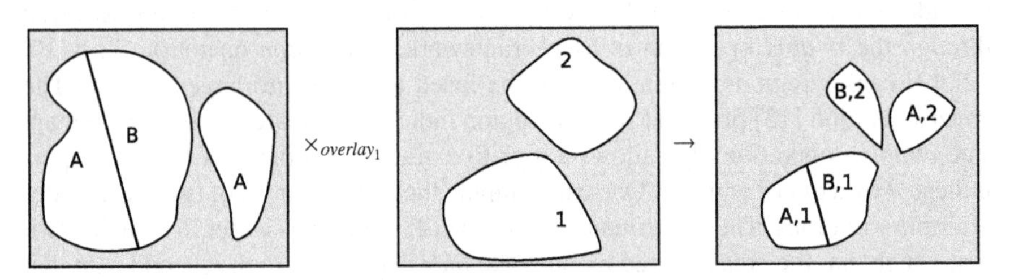

Fig. 5.1 A spatial partition π of type A, a spatial partition σ of type B, and the result of the operation $overlay_1(\pi, \sigma)$. The exterior of each spatial partition is shaded

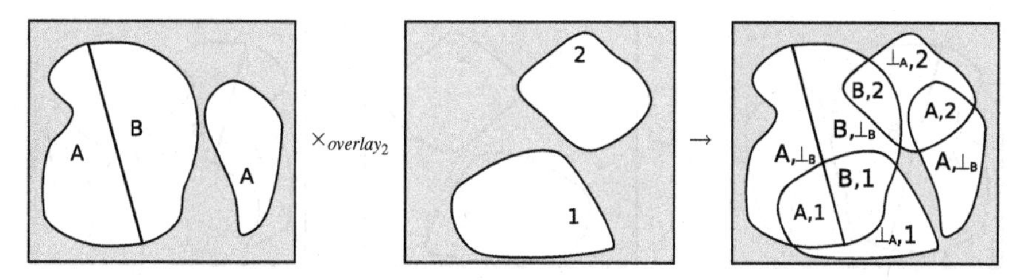

Fig. 5.2 A spatial partition π of type A, a spatial partition σ of type B, and the result of the operation $overlay_2(\pi, \sigma)$. The exterior of each spatial partition is shaded

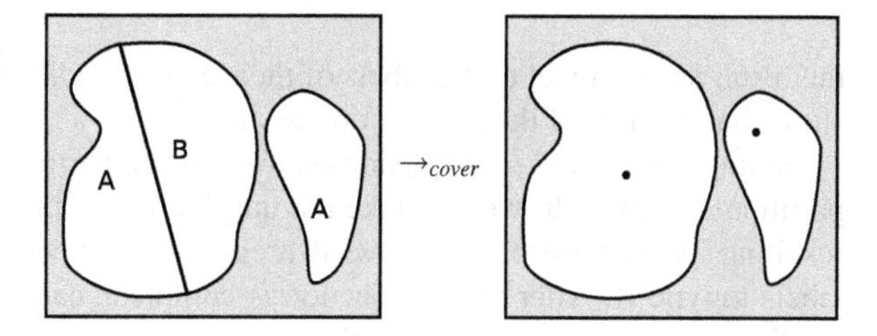

Fig. 5.3 A spatial partition π of type A and the result of the operation $cover(\pi)$. The exterior of each spatial partition is shaded

5.3 Cover

The *cover* operation takes a single spatial partition π and returns a spatial partition containing a single region that is identical to the geometric union of all regions in π. This operation can be achieved quite simply with a relabel operation; effectively, any region label that is not the exterior label is relabeled to an identical label. In effect, this will fuse all the regions together. We use the $\bullet$ label as the label of the fused region. Figure 5.3 depicts an example of the cover operation.

$$\begin{aligned} &cover : [A] \rightarrow [U] \\ &cover(\pi) := relabel(\pi, \lambda x : A.\ \textit{if } x \neq \perp_A \textit{ then } \bullet) \end{aligned} \tag{5.3}$$

5.4 Difference

There are many possible constructions of difference operations on spatial partitions. Difference operations can be symmetric or non-symmetric and can combine labels from input partitions or not. We examine each of these possibilities.

We begin with the non-symmetric version of the difference operator, which we denote $l - diff$. Under the $l - diff$ operation, two spatial partitions, π and σ, are provided and the portions of π that are covered by regions in σ are removed from π in the result. Because the result of the operation contains only regions from pi, the type of the result is identical to the type of π; in other words, the labels of regions in

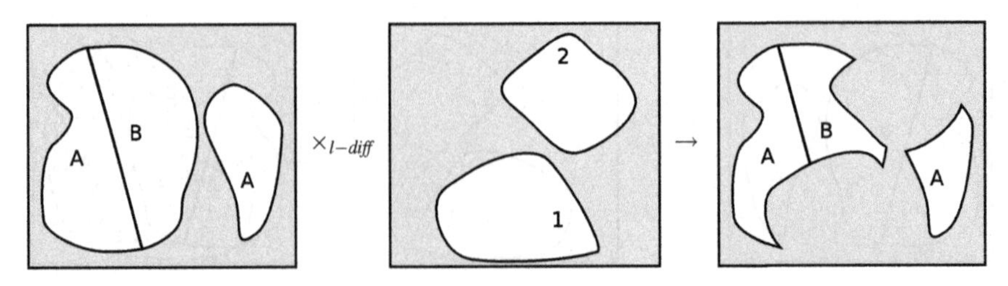

Fig. 5.4 A spatial partition π of type A, a spatial partition σ of type B, and the result of the operation $l - \mathit{diff}(\pi, \sigma)$. The exterior of each spatial partition is shaded

the result are identical to the labels of their corresponding regions in pi. Figure 5.4 shows an example of the $l - \mathit{diff}$ operation.

To implement the $l - \mathit{diff}$ operation over spatial partition π of type A and spatial partition σ of type B, we will take the intersection of the π and σ then relabel the resulting spatial partition to remove all areas covered by σ, and convert all remaining labels to type A. After the intersection is complete, each label will be a pair such the first element is of type A and the second element is of type B; therefore, every region that lies in the exterior of σ will keep the element of type A as its label, all other regions are converted to the $\bot_A$ label:

$$
\begin{aligned}
&l - \mathit{diff} : [A] \times [B] \rightarrow [A] \\
&l - \mathit{diff}(\pi, \sigma) := relabel(\ intersection(\pi, \sigma), \\
&\qquad \lambda(a, b) : A \times B. \\
&\qquad\quad \mathit{if}\ b = \bot_B\ then\ a \\
&\qquad\quad else\ \bot_A)
\end{aligned}
\tag{5.4}
$$

The $r - \mathit{diff}$ operation is equivalent to $l - \mathit{diff}$ with the arguments reversed:

$$
\begin{aligned}
&r - \mathit{diff} : [A] \times [B] \rightarrow [A] \\
&r - \mathit{diff}(\pi, \sigma) := l - \mathit{diff}(\sigma, \pi)
\end{aligned}
\tag{5.5}
$$

The symmetric difference operation, denoted *diff*, is similar to a geometric difference operation. The result of the *diff* operation over a spatial partition π of type A and a spatial partition σ of type B will be a spatial partition containing the areas covered by a region in only a single input partition, and will have type $A \times B$. Figure 5.5 depicts and example. *diff* is constructed by computing the intersection of the results of the $l - \mathit{diff}$ and $r - \mathit{diff}$ operations of the input partitions:

$$
\begin{aligned}
&\mathit{diff} : [A] \times [B] \rightarrow [A \times B] \\
&\mathit{diff}(\pi, \sigma) := intersection(l - \mathit{diff}(\pi, \sigma), r - \mathit{diff}(\pi, \sigma))
\end{aligned}
\tag{5.6}
$$

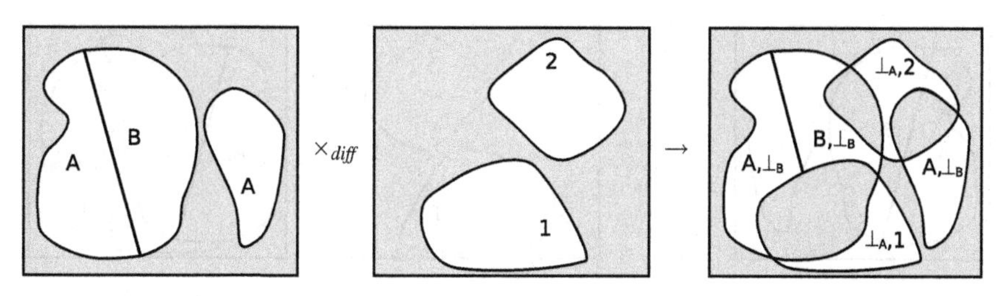

Fig. 5.5 A spatial partition π of type A, a spatial partition σ of type B, and the result of the operation $diff(\pi, \sigma)$. The exterior of each spatial partition is shaded

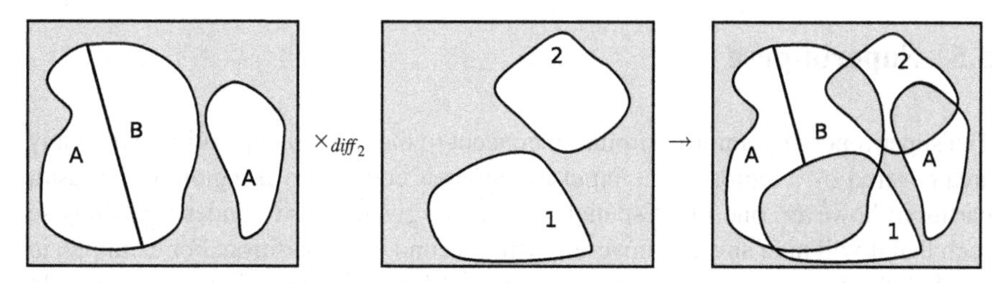

Fig. 5.6 A spatial partition π of type A, a spatial partition σ of type A, and the result of the operation $diff_2(\pi, \sigma)$. The exterior of each spatial partition is shaded

By definition, the regions in the spatial partition resulting from the *diff* operation will cover areas that are covered by a region in only a single spatial partition used as input to the operation; this implies that the label of every region in the resulting spatial partition will be a pair such that at least one element of the pair will be an exterior label of an input partition. Thus, if the types of the input partitions are identical, it is possible to alter the labels of the result partition such that the type of the output partition is identical to the type of both input partitions. We denote this operation $diff_2$. Figure 5.6 shows an example.

The $diff_2$ operation will be implemented by relabeling the intersection of the input regions. For the example in Fig. 5.6, the result of the intersection will be regions labeled with pairs of type $A \times A$. To construct the difference of the partitions, we simply convert labels back to type A. If a label pair contains both exterior labels, we convert it to $\bot_A$. If a label contains two labels that are not exterior labels, we convert it to $\bot_A$. This leaves only labels where exactly one element in the pair is the exterior label; for those labels, we simply keep the non-exterior label. Formally:

$$
\begin{aligned}
& diff_2 : [A] \times [A] \rightarrow [A] \\
& diff_2(\pi, \sigma) := relabel(\ intersection(\pi, \sigma), \\
& \qquad \lambda(a, b) : A \times A. \\
& \qquad\quad if\ a \neq \bot_A \wedge b \neq \bot_A\ then\ \bot_A \\
& \qquad\quad else\ if\ a = \bot_A \wedge b = \bot_A\ then\ \bot_A \\
& \qquad\quad else\ if\ b = \bot_A\ then\ a \\
& \qquad\quad else\ if\ a = \bot_A,\ then\ b)
\end{aligned}
\tag{5.7}
$$

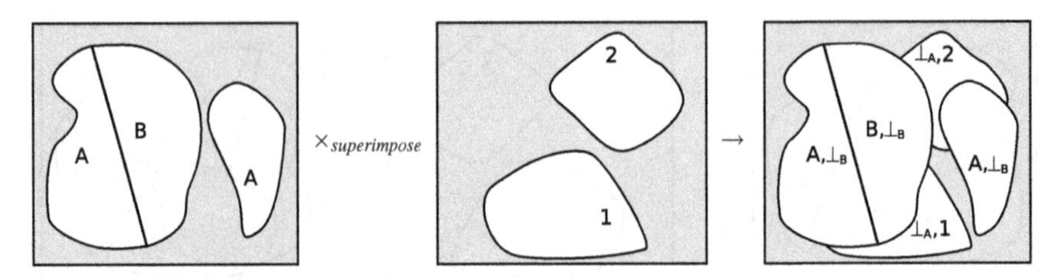

Fig. 5.7 A spatial partition π of type A, a spatial partition σ of type B, and the result of the operation $superimpose(\pi, \sigma)$. The exterior of each spatial partition is shaded

5.5 Superimpose

The superimpose operator is similar in concept to the overlay$_2$ operation in that any area covered by a region in the input partitions is covered by a region in the result partition; however, one input spatial partition is given priority under superimpose such that it will hide any ares covered by the second input partition. For example, in Fig. 5.7, the regions in π remain unchanged in the superimposition of π and σ. In effect, some areas of σ are hidden by regions in π.

The superimpose operation is constructed by relabeling the intersection of the input spatial partitions. The labels in the result of the intersection will be pairs. Any pair that has two non-exterior labels is changed such the the second item (the item of type B for the case in Fig. 5.7), is changed to the exterior label of the appropriate type. Thus, the regions from the first input partition are unchanged in the result of a superimposition.

$$\begin{aligned} &superimpose : [A] \times [B] \rightarrow [A \times B] \\ &superimpose(\pi, \sigma) := relabel(\ intersection(\pi, \sigma), \\ &\qquad \lambda(a, b) : A \times B. \\ &\qquad\quad if\ a \neq \perp_A \wedge b \neq \perp_B\ then\ (a, \perp_B)) \end{aligned} \tag{5.8}$$

5.6 Window

The window operation allows a user keep only the regions in a spatial partition that are non-disjoint with a user-defined query window. The query window is a spatial partition itself. For example, Fig. 5.8 shows a spatial partition, a window, and the result containing only regions that are non-disjoint to the query window.

The window operation is different than any previous operation because we cannot determine if two regions in a spatial partition are non-disjoint by simply examining the region labels in isolation. For example, if a region meets a region in the query window, then the labels of those two regions will together form a boundary label in the intersection of the spatial partition and query window; however, we must be able to examine the entire boundary label to learn of the meet. The labels of the

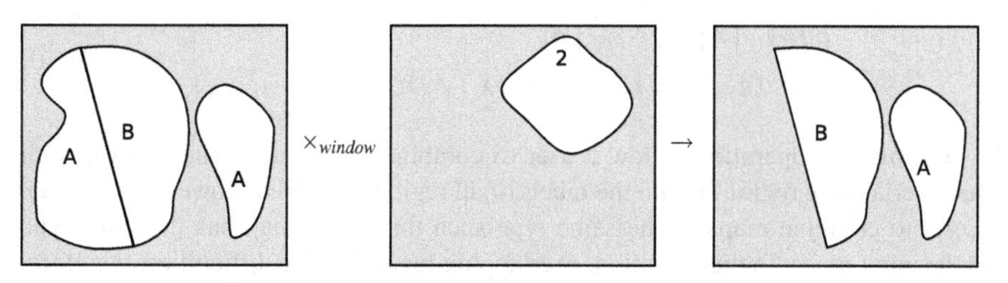

Fig. 5.8 A spatial partition π of type A, a spatial partition σ of type B, and the result of the operation $window(\pi, \sigma)$. The exterior of each spatial partition is shaded

intersection of a spatial partition and a query window do contain all the required information, therefore, we must be able to examine the range of a spatial partition into order to specify the window operation.

The window operation first computes τ: the intersection of a spatial partition π and a second spatial partition σ defining the window. Regions in π that meet a region in σ will cause a border label in τ to contain two labels, each defined as pairs, such that each pair respectively contains: (i) a non-exterior label from π and the exterior label from σ, and (ii) the exterior label from π and a non-exterior label from σ. Because we are interested in keeping regions from π, the element that is a non-exterior label from π will be added to a list l_1. If a region from π and σ overlap, a label will exist in the range that is a pair containing non-exterior elements from both π and σ. In such a case, the element from π is added to a list L_2. L_1 and L_2 then contain the labels of all regions from π that are non-disjoint from some region in σ. The final step is to relabel τ to keep only the regions whose label is in those lists, and convert the labels of those regions to the type of π:

$$
\begin{aligned}
&window : [A] \times [B] \rightarrow [A] \\
&window(\pi, \sigma) := \tau \leftarrow intersection(\pi, \sigma) \\
&\qquad L_1 \leftarrow \{a | (a \neq \bot_A, b \neq \bot_B) \in S \in rng(\tau)\} \\
&\qquad L_2 \leftarrow \{a | (a \neq \bot_A, \bot_B), (\bot_A, b \neq \bot_B) \in S \in rng(\tau)\} \\
&\qquad relabel(\tau, \lambda(a, b) : A \times B.\ if\ a \in L_1 \cup L_2\ then\ a\ else\ \bot_A)
\end{aligned}
\tag{5.9}
$$

5.7 Operations to Support Layering

The ability to extract elements from a map into separate *layers* and to combine those layers into a single map layer is sometimes useful in map analysis. Again, we are able to specify such operations using the power operations.

The select operation allows a user to extract regions from a map by providing their labels. The select operation can be combined with a refine operation if single faces of regions are required. The select operation takes a spatial partition and a set of labels of regions desired in the result map. To select regions from a map, we relabel all non-requested regions to the exterior label:

$$\begin{aligned} &select : [A] \times 2^A \rightarrow [A] \\ &select(\pi, S) := relabel(\pi, \lambda x : A.\ if\ x \notin S\ then\ \perp_A) \end{aligned} \tag{5.10}$$

The overlay operations allow a user to combine maps into a single layer such that overlapping regions retain the labels of all regions involved; however, it is often useful to combine maps of the same type such that result map has the same type as the input maps. In this situation, overlapping regions cause difficulties; therefore, we require an operation that retains the type of the input maps, and removes areas covered by regions in more than one input map. The $diff_2$ provided the required functionality, but its name is slightly confusing in the layering context. Therefore, we provide the alternate name *union*:

$$\begin{aligned} &union : [A] \times [A] \rightarrow [A] \\ &union(\pi, \sigma) := diff_2(\pi, \sigma) \end{aligned} \tag{5.11}$$

We can then use *select* and *union* to define more general layering functionality. The *layer* operation takes a set of region labels and a spatial partition and will create a separate map for every region in the partition whose label is in the set.

$$\begin{aligned} &layer : [A] \times 2^A \rightarrow 2^{[A]} \\ &layer(\pi, S) := \{select(\pi, s \in S)\} \end{aligned} \tag{5.12}$$

The collapse operation will take a set of spatial partitions and computer their union, returning a single map:

$$\begin{aligned} &collapse : 2^{[A]} \rightarrow [A] \\ &collapse(P) := union_{\pi \in P} \pi \end{aligned} \tag{5.13}$$

Let π be a spatial partition of type A, π is equivalent to collapsing all layers of π:

$$\pi = collapse(layer(\pi, \{x | x \in rng(\pi) \wedge |x| = 1\})) \tag{5.14}$$

5.8 Summary

In this chapter, we have shown how to express map operations from the literature in the Map Framework. All operations are constructed using the power operations, thus type closure under spatial partitions is maintained. In the following chapter, we expand upon this set of operations to explore more advanced operations using notions of connectivity and joins.

References

1. Berry, J.K.: Fundamental operations in computer-assisted map analysis. International Journal of Geographical Information System **1**(2), 119–136 (1987)
2. Chan, E.P., Zhu, R.: Ql/g-a query language for geometric data bases. In: 1st Int. Conf. on GIS in Urban and Environmental Planning, pp. 271–286. Citeseer (1996)
3. Dangermond, J.: A classification of software components commonly used in geographic information systems. Design and implementation of computer-based geographic information systems pp. 70–91 (1983)
4. Erwig, M., Schneider, M.: Partition and conquer. In: Spatial Information Theory A Theoretical Basis for GIS, pp. 389–407. Springer (1997)
5. Erwig, M., Schneider, M.: Formalization of advanced map operations. In: 9th Int. Symp. on Spatial Data Handling, vol. 8, pp. 3–17 (2000)
6. Frank, A.U.: Overlay Processing in Spatial Information System. University of Maine (1987)
7. Güting, R.H.: Geo-relational algebra: A model and query language for geometric database systems. Springer (1988)
8. Güting, R.H., Schneider, M.: Realm-based spatial data types: the rose algebra. The VLDB Journal The International Journal on Very Large Data Bases **4**(2), 243–286 (1995)
9. Huang, Z., Svensson, P., Hauska, H.: Solving spatial analysis problems with geosal, a spatial query language. In: SSDBM, pp. 1–17 (1992)
10. Kriegel, H.P., Brinkhoff, T., Schneider, R.: The combination of spatial access methods and computational geometry in geographic database systems. In: Advances in Spatial Databases, pp. 5–21. Springer (1991)
11. McKenney, M., Schneider, M.: Advanced operations for maps in spatial databases. Progress in Spatial Data Handling pp. 495–510 (2006)
12. Schneider, M.: Spatial data types for database systems(finite resolution geometry for geographic information systems). Lecture notes in computer science (1997)
13. Scholl, M., Voisard, A.: Thematic map modeling. In: Design and Implementation of Large Spatial Databases, pp. 167–190. Springer (1989)
14. Tomlin, C.D.: Geographic information systems and cartographic modeling. 526.0285 T659. Prentice Hall (1990)
15. Valenzuela, C.: Data analysis and modeling. Remote Sensing and Geographical Information Systems for Resource Management in Developing Countries pp. 335–348 (1991)

References

Chapter 6
Extended Operations Over Maps

Chapter 4 described the foundational (power) map operations and Chap. 5 showed how to define common operations over maps using the power operations. In this chapter, we develop a class of *map join* operations that allows a user to overlay maps while keeping regions that satisfy certain predicates. We then consider operations involving connectivity concepts, and extend this class of operations by defining new, more complex operations that take advantage of the connectivity properties of maps. The operations developed in this chapter were originally proposed in [3].

Although the operations in this chapter involve complex notions of connectivity and joining maps, they are still able to be expressed in terms of the power operations. In fact, most of the complexity lies in predicates that must examine the structure of a map in order to provide data to relabel operations that will keep or discard regions in maps. Thus, these operations further indicate the importance of the power operations.

6.1 Motivating Query Scenarios

In this section, we consider map operations from the user's perspective. We assume that a user will have access to a system that provides map access and the ability to perform operations over maps (i.e., a map based GIS system or a map database). A series of scenarios is presented in which the user requires certain information from the maps in the system. We then specify, in English sentences, the specific results required by the user in such scenarios. By examining these specifications in later sections, we discover the need for new map operations.

Throughout this chapter, we study map operations as they pertain to specific examples; however, the operations presented are general operations that are relevant to situations other than the ones we show here. In fact, we find that we can define only a few powerful operations that can be generalized to apply to many cases through the use of different operands. We merely present example scenarios to motivate the utility of the operations and to help explain their semantics. For the sake of clarity,

M. McKenney and M. Schneider, *Map Framework*,
DOI 10.1007/978-3-319-46766-5_6

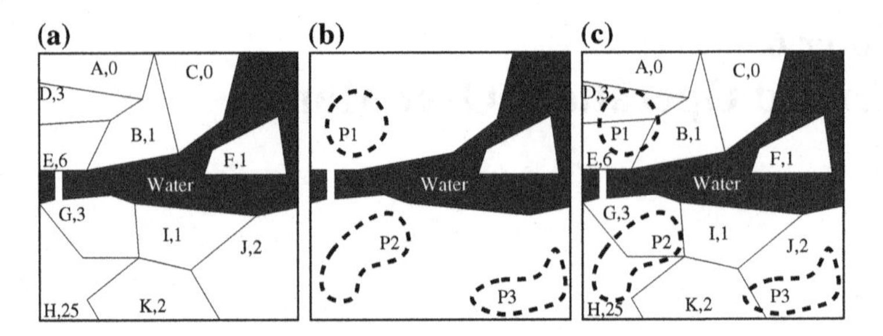

Fig. 6.1 Figure **a** shows a map of counties. Figure **b** shows areas of high population density. Figure **c** shows (**b**) superimposed over (**a**)

we will only introduce three sample maps, shown in Fig. 6.1, against which queries requiring map operations will be posed. The first map in Fig. 6.1a depicts counties around a body of water. Each county is labeled with an identifying letter, and the number of flu vaccine units currently in that county. County G contains a bridge over the water which is represented by an extension of G that touches county E. Figure 6.1b contains regions enclosed by dashed lines which represent areas of high population density. The third map, shown in Fig. 6.1c shows the areas of high population density superimposed over the map of counties. Operations will be performed over the first two maps, the third is shown for reference. Note that the water shown in the figures is not part of the actual map stored in the database, but is shown merely to provide additional context.

6.1.1 Map Joining Scenarios

For the remainder of this chapter, we will assume that the flu virus has been detected in all of the areas of high population density represented in Fig. 6.1b. Frequently, supplies such as flu vaccines are distributed through local governments, such as county governments in the United States. Therefore, state officials need to know which counties contain the infected areas of high population density so that additional vaccinations may be sent there. The officials can obtain this information by combining information from the population density map and the county map to calculate a map containing only counties that overlap a region of high population density (the term 'overlap' indicates the topological predicate *overlap* between two complex regions. Chapter 7 covers topological predicates in more detail.). This type of query cannot be calculated using existing map operations. Therefore, a new operation is required that allows a user to create a map consisting of complete faces of regions from two source maps that satisfy a given predicate. We call this type of operation a *map join*. The result of such a scenario is shown in Fig. 6.2a. Only regions from the county map that overlap a region in the population density map remain in the result map. Note is that this particular example represents a one-sided join, i.e., only regions from one of the argument maps are being returned. Two sided joins are also possible, in which regions from both argument maps that satisfy the predicate are shown in the result map.

Query 1 *Return the map consisting of all counties from the county map that overlap a region in the population density map.*

The map join operation is parametrized by a user supplied predicate; thus, different queries can be posed over data by simply changing the predicate used in the join. For example, instead of finding counties that overlap a region of high population density, the official might want to know which counties meet a region of high population density. Such a query would identify counties in which the flu is likely to spread. To express this query, we simple replace the word 'overlap' in Query 1 with the word 'meet'. The resulting map is shown in Fig. 6.2b.

A second map join scenario considers the connectivity information which is inherently represented in a map's structure. Suppose that the particular strain of the flu that the authorities are tracking is highly contagious. In this case, the authorities need to know all counties that are connected through one or more counties to a county that overlaps an infected region of high population density. Such a map (depicted in Fig. 6.2c) represents all possible counties in danger of becoming infected.

Query 2 *Return the map consisting of all counties that overlap a region of high population density, and all counties that are connected through one or more counties to county that overlaps a high population density region.*

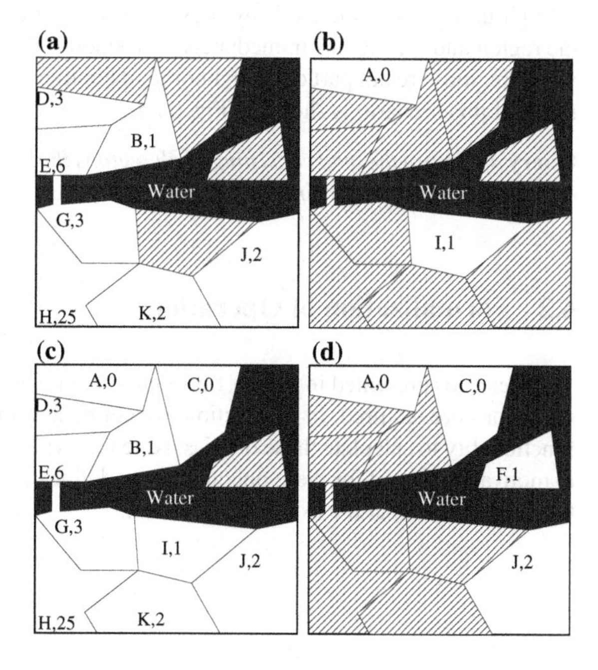

Fig. 6.2 The resulting partitions from the sample queries. *Shaded regions* and the water areas are not included in the actual map, but are shown for reference

6.1.2 Complex Connectivity Scenarios

One of the advantages of using maps as a data type is that topological and connectivity information is inherently stored in the map in a visual form. While some work has been done on identifying and defining operations that utilize connectivity information, the extent to which such information can be used has not been explored. Here we consider scenarios in which connectivity calculations are required beyond those offered by existing map operations.

A more complex extension of the basic notion of connectivity is the idea of connectivity through a specific construct. For example, county H in Fig. 6.1a contains far more units of flu vaccine than the other counties. Assume that a large number of transport vehicles are stationed in county J. A useful query in this situation is to find out if county C is connected to county J through county H. If the connection exists, then the transport vehicles can drive to county H, pick up vaccines, and deliver them to county C which has no vaccines. There are two possible versions of such a query, one that returns the map of counties that connect C and J through H, and one that simply returns a true or a false value. A query that calculates the latter would be stated:

Query 3 *Are counties C and J connected through county H?*

Instead of transporting vaccines around the map, the authorities determine that if the sum of vaccine units in a county and all of its neighboring counties is 6 or greater, that county has access to enough vaccines within a short distance to effectively combat the spread of the virus. A query to calculate which regions have enough units of vaccine close by uses the notion of a region's neighborhood, i.e., the region and the regions immediately surrounding it. The authorities need to know which regions are not part of a neighborhood containing at least 6 vaccine units. The result of this scenario is shown in Fig. 6.2d.

Query 4 *Return the map consisting of all regions that are not part of a neighborhood containing at least 6 vaccine units.*

6.2 Formalization of Operations

The scenarios presented in Sect. 6.1 provide specific instances in which new types of operations are required. In this section, we define new operations which provide the functionality to calculate those queries. Note that while we used the example queries to motivate the need for these operations, we define operations to satisfy the general types of queries that the examples represent.

6.2.1 Join Operations

We define the map join in two parts. First, we must define a method by which regions in one of the argument partitions that satisfy the join constraint can be collected. We achieve this in the *collect* operation which takes two partitions, π of type A and σ of type B, and a predicate P (which takes two complex regions), and relabels any region from π that does not satisfy the predicate with at least one region from σ to $\bot$.

$$\begin{aligned} &collect : [A] \times [B] \times (2^{\mathbb{R}^2} \times 2^{\mathbb{R}^2} \to \mathbb{B}) \to [A] \\ &collect\ (\pi, \sigma, P) := relabel(\pi, \lambda x : A.\ \text{if } \nexists y \in B | P(\overline{\pi^{-1}(x)}, \overline{\sigma^{-1}(y)}) \\ &\qquad\qquad\qquad\qquad\qquad\qquad\qquad \text{then } \bot_A) \end{aligned} \tag{6.1}$$

The join operation can then be defined using the operations *intersection* and *collect*. Since a single region may consist of multiple faces within the partition, we use the refine operation to break such regions so that each face is considered a single region. Because refine extends each region label by appending an integer, we will also use a relabel operation after each collect operation that simply removes the appended integer from each region. We denote this operation *truncate*.

$$\begin{aligned} &truncate : [A] \to [B] \\ &truncate(\pi) := relabel(\pi, \lambda(a, b) : A.\ \text{then } a) \end{aligned} \tag{6.2}$$

Given two partitions π and σ and a predicate P, we define the *left-join* and *right-join* operations which create a result partition consisting of regions from the first and second argument partitions, respectively (i.e., they are one-sided joins):

$$\begin{aligned} &left\text{-}join : [A] \times [B] \times (2^{\mathbb{R}^2} \times 2^{\mathbb{R}^2} \to \mathbb{B}) \longrightarrow [A] \\ &left\text{-}join(\pi, \sigma, P) := truncate(collect(refine(\pi), refine(\sigma), P)) \\ &right\text{-}join : [A] \times [B] \times (2^{\mathbb{R}^2} \times 2^{\mathbb{R}^2} \to \mathbb{B}) \longrightarrow [B] \\ &right\text{-}join(\pi, \sigma, P) := truncate(collect(refine(\sigma), refine(\pi), P)) \end{aligned} \tag{6.3}$$

The two-sided join, which we denote simply as *join*, returns the partition consisting of all regions from both argument partitions π and σ that satisfy a given predicate P.

$$\begin{aligned} &join : [A] \times [B] \times (2^{\mathbb{R}^2} \times 2^{\mathbb{R}^2} \to \mathbb{B}) \longrightarrow [A \times B] \\ &join(\pi, \sigma, P) := intersection(left\text{-}join(\pi, \sigma, P), right\text{-}join(\pi, \sigma, P)) \end{aligned} \tag{6.4}$$

6.2.2 Basic Connectivity Using Joins

The map join operations in the previous section are general enough to calculate any join that can be expressed based on a predicate that is satisfied between regions from opposing partitions. However, Query 2 introduces the idea of incorporating connectivity with a map join operation. The result partition for this query must contain the regions that satisfy a left-join operation using the overlap predicate, and regions from the left partition that are connected (through one or more regions) to the result of the left-join. For example, to compute Fig. 6.2c, Query 2 must first compute the partition in Fig. 6.2a using a left-join with the overlap predicate, then add more regions from the county map to the result if they satisfy a second predicate with another region in the county map that is part of the left-join, namely if they are connected to a region in the result of the join. We denote this new operation the *extended-join* between two partitions.

We construct the extended-left-join to be parameterized based on user predicates. The procedure takes multiple steps. First, we compute the left-join of two partitions where regions satisfy a given predicate P. In the case of Query 2, we are interested in regions from the left partition that overlap a region from the right partition. We then use the *difference* operator (Chap. 5 and [1]) to compute the regions from the original left partition that are not in the left-join (the difference operator takes two partitions and returns the first partition minus any area that is overlapped by the second partition). Out of those regions, we collect the regions that satisfy a second predicate Q with some region in the result of the left-join. For the purposes of Query 2, we must use a predicate that returns *True* if a region is connected through one or more regions to a region in the result of the left-join. If we only want regions directly connected to the result of the left-join, a *meet* topological predicate may be used. We denote the portion of the operation in which predicate Q is involved as the *extended portion of the join.*

Given two partitions, π and σ, a predicate P for the join, and a predicate Q for the extended portion of the join, we first calculate the left-join (right-join) based on predicate P. We then include, using the operations intersection and collect, any regions in the difference of π (σ) and the result of the left-join (right-join) that satisfy the predicate Q with a region in the left-join (right-join).

$$
\begin{aligned}
&\mathit{extended{-}left{-}join} : [A] \times [B] \times (2^{\mathbb{R}^2} \times 2^{\mathbb{R}^2} \rightarrow \mathbb{B}) \times (2^{\mathbb{R}^2} \times 2^{\mathbb{R}^2} \rightarrow \mathbb{B}) \longrightarrow [A] \\
&\mathit{extended{-}left{-}join}(\pi, \sigma, P, Q) := \\
&\quad \mathit{union}(\\
&\qquad \mathit{collect}(\\
&\qquad\quad l - \mathit{diff}(\pi, \mathit{left{-}join}(\pi, \sigma, P)), \\
&\qquad\quad \mathit{left{-}join}(\pi, \sigma, P), Q), \\
&\qquad \mathit{left{-}join}(\pi, \sigma, P) \\
&\quad)
\end{aligned}
\tag{6.5}
$$

$$
\begin{array}{l}
\textit{extended-right-join} : [A] \times [B] \times (2^{\mathbb{R}^2} \times 2^{\mathbb{R}^2} \to \mathbb{B}) \times (2^{\mathbb{R}^2} \times 2^{\mathbb{R}^2} \to \mathbb{B}) \longrightarrow [B] \\
\textit{extended-right-join}(\pi, \sigma, P, Q) := \\
\quad \textit{union}(\\
\quad\quad \textit{collect}(\\
\quad\quad\quad \textit{l} - \textit{diff}(\sigma, \textit{right-join}(\pi, \sigma, P)), \\
\quad\quad\quad \textit{right-join}(\pi, \sigma, P), Q), \\
\quad\quad \textit{right-join}(\pi, \sigma, P) \\
\quad)
\end{array} \tag{6.6}
$$

The two-sided extended join can then be defined as the intersection of the left and right extended joins. Because the second predicate used by the left and right extended joins is evaluated between regions of the first and second argument partition, respectively, we pass two additional predicates to the two-sided extended join, one which is used by the first argument partition, and the other which is used by the second argument partition.

$$
\begin{array}{l}
\textit{extended-join} : [A] \times [B] \times (2^{\mathbb{R}^2} \times 2^{\mathbb{R}^2} \to \mathbb{B}) \times (2^{\mathbb{R}^2} \times 2^{\mathbb{R}^2} \to \mathbb{B}) \\
\qquad\qquad\qquad \times (2^{\mathbb{R}^2} \times 2^{\mathbb{R}^2} \to \mathbb{B}) \longrightarrow [A \times B] \\
\textit{extended-join}(\pi, \sigma, P, Q, R) := \textit{intersection}(\textit{extended-left-join}(\pi, \sigma, P, Q), \\
\qquad\qquad\qquad\qquad\qquad\qquad \textit{extended-right-join}(\pi, \sigma, P, R))
\end{array} \tag{6.7}
$$

Clearly, topological predicates between regions are appropriate in the extended join operations. Query 2 can be computed using an extended left join operation; however, to complete the query, a predicate is required that takes two regions in a partition and returns a value of *True* if those regions are connected, and returns a value of false otherwise. We define the *connected* predicate that takes two regions r and s each belonging to a separate argument partition. This predicate returns a value of *True* if there exists a chain of regions from r to s such that each region in the chain is not disjoint with its neighbors in the chain (i.e., the regions share at least one common point). Furthermore, each region in the chain can be from either argument partition. For example, if the two argument partitions were the county map and the population density map, then the chain J, P3, K, H, P2 connects regions J and P2. Given two regions r and s from partitions π and σ respectively, we define the connected predicate as follows:

$$
\begin{array}{l}
\textit{connected} : [A] \times A \times [B] \times B \longrightarrow \mathbb{B} \\
\textit{connected}(\pi, r, \sigma, s) := \\
\begin{cases}
\textit{True} & \text{if } \exists t_1 \ldots t_n | t_1 = r \wedge t_n = s \wedge \forall 1 \leq i < n : \\
& ((t_i, t_{i+1} \in A - \{\bot_A\} \wedge \neg \textit{disjoint}(\overline{\pi^{-1}(t_i)}, \overline{\pi^{-1}(t_{i+1})})) \\
& \vee (t_i, t_{i+1} \in B - \{\bot_B\} \wedge \neg \textit{disjoint}(\overline{\sigma^{-1}(t_i)}, \overline{\sigma^{-1}(t_{i+1})})) \\
& \vee (t_i \in A - \{\bot_A\} \wedge t_{i+1} \in B - \{\bot_B\} \wedge \neg \textit{disjoint}(\overline{\pi^{-1}(t_i)}, \overline{\sigma^{-1}(t_{i+1})})) \\
& \vee (t_i \in B - \{\bot_B\} \wedge t_{i+1} \in A - \{\bot_A\} \wedge \neg \textit{disjoint}(\overline{\sigma^{-1}(t_i)}, \overline{\pi^{-1}(t_{i+1})}))) \\
\textit{False} & \text{otherwise}
\end{cases}
\end{array} \tag{6.8}
$$

The purpose of Query 2 is to find all counties in the county map that the flu is likely to spread to in the future based on the connectivity of a county to a county which already has flu infections present in it. However, no limit is set on the number of counties in a connected chain. Instead, the user may wish to calculate an extended join containing counties that are connected to an infected county by only one or two counties. This query allows the user to see all infected counties, and all counties which are likely to experience flu infections in the near future. To define such a predicate, we introduce the concept of degree of connectivity between two regions.

The *degree of connectivity* between two regions indicates the minimum number of regions, not counting the source and destination regions, that are needed in a particular partition to connect a source region and a destination region. If two regions with labels r and s in a partition π are not connected, then their connectivity degree is -1. If they are connected directly (meaning they are not disjoint, or they share at least one common point), then their connectivity degree is 0. If a single region connects them, then their connectivity degree is 1, etc. Note that some regions may be connected by more than one path of regions, in which case the connectivity degree is the degree of the minimal path. For example, counties H and K in the county map have a connectivity degree of zero even though they are connected by paths *H, K* and *H, G, I, J, K*.

$$
\begin{array}{l}
\textit{degree-connected} : [A] \times A \times A \longrightarrow \mathbb{N} \\
\textit{degree-connected}(\pi, r, s) := \\
\begin{cases}
-1 & \text{if } \neg \textit{connected}(\pi, r, \pi, s) \\
0 & \text{if } \textit{connected}(\pi, r, \pi, s) \wedge \neg \textit{disjoint}(\overline{\pi^{-1}(r)}, \overline{\pi^{-1}(s)}) \\
n & \text{if } \exists t_1, \ldots, t_n \in A - \{\bot_A\} | \pi^{-1}(t_1), \ldots, \pi^{-1}(t_n) \in \rho(\pi) \wedge \textit{connected}(\pi, r, \pi, s) \\
 & \wedge \neg \textit{disjoint}(\overline{\pi^{-1}(r)}, \overline{\pi^{-1}(t_1)}) \wedge \ldots \wedge \neg \textit{disjoint}(\overline{\pi^{-1}(t_n)}, \overline{\pi^{-1}(s)}) \\
 & \text{where } n \text{ is minimal}
\end{cases}
\end{array}
\tag{6.9}
$$

6.2.3 Formalization of Complex Connectivity Operations

The notions of connectivity defined in the previous section build upon the idea of determining if two regions in a map are connected. In this section, we define new, more advanced connectivity operations.

Query 3 asks if two regions are connected through a specific region in the map. In order to determine whether two regions are connected through a specified region, we define the operation *degree-connected-through*, which takes a partition, a source and destination region label, and a region label which identifies a region that the connection must pass through. The degree of connectivity returned is the minimum number of regions, including the region required to be part of the path, that are needed to connect the source and destination regions, or -1 if such a connection does not exist:

$$
\begin{aligned}
&\textit{degree-connected-through} : [A] \times A \times A \times A \longrightarrow \mathbb{N}\\
&\textit{degree-connected-through}(\pi, r, s, q) :=\\
&\quad \begin{cases}
-1 & \text{if } \neg \textit{connected}(\pi, r, \pi, q)\\
 & \quad \vee \neg \textit{connected}(\pi, q, \pi, s)\\
0 & \text{if } \textit{degree-connected}(\pi, r, s) = 0\\
 & \quad \wedge (r = q \vee s = q)\\
\textit{degree-connected}(\pi, r, q) & \\
\quad + \textit{degree-connected}(\pi, s, q) & \text{if } (r = q \vee s = q)\\
\textit{degree-connected}(\pi, r, q) & \\
\quad + \textit{degree-connected}(\pi, s, q) + 1 & \text{otherwise}
\end{cases}
\end{aligned}
\tag{6.10}
$$

It is sometimes useful to calculate the partition containing the connection of two regions. We use the *connect* operation to provide a partition containing a source and destination region, and the minimum number of regions that connect them. We first define the *connected-set* operation to return the labels of the regions that connect two argument regions:

$$
\begin{aligned}
&\textit{connected-set} : [A] \times A \times A \rightarrow [A]\\
&\textit{connected-set}(\pi, r, s) :=\\
&\quad \{t_1, \ldots, t_n | n = \textit{degree-connected}(\pi, r, s) + 2 \wedge t_1 = r \wedge t_n = s \wedge\\
&\quad (\forall 1 \leq i \leq n : t_i \neq \bot_A) \wedge\\
&\quad (\forall 1 \leq j \leq n : \pi^{-1}(t_j) \in \underline{\rho(\pi)}) \wedge\\
&\quad (\forall 1 \leq k < n : \neg \textit{disjoint}(\overline{\pi^{-1}(t_k)}, \overline{\pi^{-1}(t_{k+1})}))\}
\end{aligned}
\tag{6.11}
$$

The *connect* operation [2] simply relabels any regions not in the connected-set to the exterior label:

$$
\begin{aligned}
&\textit{connected} : [A] \times A \times A \rightarrow [A]\\
&\textit{connected}(\pi, r, s) :=\\
&\quad \textit{relabel}(\pi, \lambda l : A. \text{ if } l \notin \textit{connected} - \textit{set}(\pi, w, x) \text{ then } \bot_A)
\end{aligned}
\tag{6.12}
$$

The *connect-through* operation calculates the partition consisting of a source and destination region, and the minimum number of regions that connect them, including a specified region. This operation can be reduced to a relabeling problem if the set of region labels contained in the resulting partition is known (i.e., the set of labels that correspond to the minimal chain of regions connecting a source and destination region). Using the degree of connectivity, we first define the operation *connected-through-set* which calculate this set of region labels. *connected-through-set* takes a spatial partition, a source region, a destination region, and the region that the connection must pass through:

$$
\begin{aligned}
&\textit{connected-through-set} : [A] \times A \times A \times A \longrightarrow 2^A \\
&\textit{connected-through-set}(\pi, r, s, q) := \\
&\quad \{t_1, \ldots, t_n | \\
&\quad n = \textit{degree-connected-through}(\pi, r, s, q) + 2 \wedge \\
&\quad t_1 = r \wedge t_n = s \wedge \\
&\quad (\exists 1 \leq i \leq n : t_i = q) \wedge \\
&\quad (\forall 1 \leq j \leq n : \pi^{-1}(t_j) \in \rho(\pi)) \wedge \\
&\quad (\forall 1 \leq k < n : \neg \textit{disjoint}(\overline{\pi^{-1}(t_k)}, \overline{\pi^{-1}(t_{k+1})})) \wedge \\
&\quad (\forall 1 \leq l \leq n : t_l \neq \bot_A)\}
\end{aligned}
\tag{6.13}
$$

The connect-through operation then relabels any region in the argument partition whose label is not in that set to the empty label:

$$
\begin{aligned}
&\textit{connect-through} : [A] \times A \times A \times A \longrightarrow [A] \\
&\textit{connect-through}(\pi, w, x, y) := \\
&\quad \textit{relabel}(\pi, \lambda l : A.\ \text{if } l \notin \textit{connected-through-set}(\pi, w, x, y) \\
&\qquad\qquad\qquad\qquad \text{then } \bot_A)
\end{aligned}
\tag{6.14}
$$

A final connectivity operation suggested in Query 4 is the neighborhood operation. In this query, the user is interested in the number of vaccine units available in the immediate neighborhood of a region (which includes that region and all regions adjacent to that region). In general, a neighborhood can be specified to include any regions within a specified degree of connectivity to a source region. For instance, the 1-neighborhood of a region consists of the source region and all regions adjacent to it. The 2-neighborhood consists of the source region and all regions with a connectivity degree of 0 or 1 to the source region. We define the neighborhood operation to take a partition π, a region label r, and an integer indicating the size of the neighborhood desired. The neighborhood is then computed by assigning the labels of all regions that are not in the neighborhood of r to $\bot$:

$$
\begin{aligned}
&\textit{neighborhood} : [A] \times A \times \mathbb{N} \longrightarrow [A] \\
&\textit{neighborhood}(\pi, r, n) := \\
&\quad \textit{relabel}(\pi, \lambda l : A.\ \text{if } \neg(\textit{degree-connected}(\pi, r, l) < n) \\
&\qquad\qquad\qquad\qquad \text{then } \bot_A)
\end{aligned}
\tag{6.15}
$$

6.3 Summary

This chapter explores more complex operations that a user may wish to compute over a map. These operations underscore the importance of the power operations, since they are constructed using the power operations and other operations defined using compositions of power operations. Furthermore, this chapter provides some querying scenarios to highlight the utility of the operations.

References

1. Erwig, M., Schneider, M.: Partition and conquer. In: Spatial Information Theory A Theoretical Basis for GIS, pp. 389–407. Springer (1997)
2. Erwig, M., Schneider, M.: Formalization of advanced map operations. In: 9th Int. Symp. on Spatial Data Handling, vol. 8, pp. 3–17 (2000)
3. McKenney, M., Schneider, M.: Advanced operations for maps in spatial databases. Progress in Spatial Data Handling pp. 495–510 (2006)

Chapter 7
Topological Relationships Between Maps

The study of topological relationships between objects in space has received much attention from a variety of research disciplines including robotics, geography, cartography, artificial intelligence, cognitive science, computer vision, image databases, spatial database systems, and geographical information systems (GIS). In the areas of databases and GIS, the motivation for formally defining topological relationships between spatial objects has been driven by a need for querying mechanisms that can filter these objects in spatial selections or spatial joins based on their topological relationships to each other, as well as a need for appropriate mechanisms to aid in spatial data analysis and spatial constraint specification.

As we have seen in earlier chapters, spatial partitions provide a data model that specifies a map in both geometric and thematic terms, and the maps inherently impose topological relationships among the regions within them; specifically, regions in spatial partitions are allowed to *meet* or be *disjoint*. The notion of topological relationships naturally precludes the consideration of thematic values in maps. Instead, the discussion of topological relationships between maps centers around the study of *map geometries*, defined as the geometric component of a spatial partition. Because only the geometric component of spatial partitions is considered, the topological relationships between spatial partitions apply to any data type that represents a partition geometry that can be created under the spatial partition model. Thus, the notion of topological relationships between maps, although developed in the context of spatial partitions, generalizes to other types of partition-based map geometries.

Consider a user who creates a map geometry modeling a swamp that is broken into regions based on pollution levels. For example, a section that is not very polluted will be represented as a region, as will be an area that is highly polluted. Once the map geometry is created, it would be interesting to query the database to find related map geometries. For instance, if someone has already completed a survey of a more specific part of the swamp, there may be map geometry in the database that consists of a more detailed view of a region in the original map geometry. Furthermore, map geometries may exist of the same swamp, but be broken into regions differently than the original map. Such map geometries may offer information as to pollution sources. Finally, it may be useful to discover which other map geometries overlap the original

M. McKenney and M. Schneider, *Map Framework*,
DOI 10.1007/978-3-319-46766-5_7

one, or are adjacent to it. A model of topological relationships between maps opens many possibilities for the querying of maps in databases.

Because map geometries are more general than complex regions, it follows that there are more possible relationships between them than there are between complex regions. Furthermore, it is unclear how such topological relationships can be implemented. Existing techniques of computing topological relationships could be extended in order to compute the new relationships, but this requires modification of database internals. If the topological relationships between map geometries could be computed based upon topological relationships between their components (i.e., their topological primitives) that correspond to existing spatial data types, then they could be directly used by existing spatial database systems.

In this chapter, we define the complete set of 49 topological relationships between pairs of map geometries that can be created under the model of spatial partitions. Furthermore, we characterize topological relationships based on their components. In this case, the components of spatial partitions are complex regions. Therefore, by characterizing the new topological relationships based on topological relationships between complex regions, we provide a method to directly use them in database systems in which topological predicates between complex regions are already implemented.

7.1 Topological Relationships Between Points, Lines, and Regions

Topological relationships indicate qualitative properties of the positions of spatial objects that are preserved under continuous transformations such as translation, rotation, and scaling. Quantitative measures such as distance or size measurements are deliberately excluded in favor of modeling notions such as connectivity, adjacency, disjointedness, inclusion, and exclusion. Attempts to model and rigorously define the relationships between certain types of spatial objects have lead to the development of two popular approaches: the *9-intersection model* [9, 10], which is developed based on point set theory and point set topology, and the *RCC model* [5–8], which utilizes spatial logic. The RCC model has found much success in spatial reasoning techniques. In the GIS area, the 9-intersection model is more commonly used.

The 9-intersection model characterizes the topological relationship between two spatial objects by evaluating the non-emptiness of the intersection between all combinations of the interior (°), boundary (∂) and exterior (⁻) of the objects involved. A unique 3 × 3 matrix, termed the *9-intersection matrix* (9IM), with values filled as illustrated in Fig. 7.1 describes the topological relationship between a pair of objects.

Fig. 7.1 The 9-intersection matrix for spatial objects A and B

$$\begin{pmatrix} A^\circ \cap B^\circ \neq \emptyset & A^\circ \cap \partial B \neq \emptyset & A^\circ \cap B^- \neq \emptyset \\ \partial A \cap B^\circ \neq \emptyset & \partial A \cap \partial B \neq \emptyset & \partial A \cap B^- \neq \emptyset \\ A^- \cap B^\circ \neq \emptyset & A^- \cap \partial B \neq \emptyset & A^- \cap B^- \neq \emptyset \end{pmatrix}$$

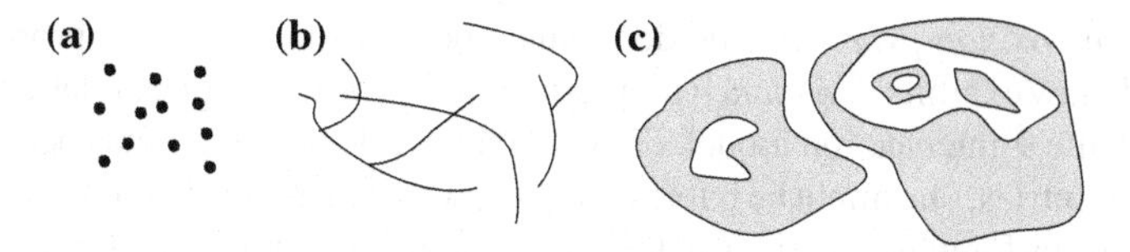

Fig. 7.2 A complex *point* (**a**), a complex *line* (**b**), and a complex region (**c**)

Fig. 7.3 The 8 topological predicates between simple regions. One object is *shaded dark* and the other light as in the *disjoint*, whereas the shared areas have the darkest shade

$\begin{pmatrix} 0 & 0 & 1 \\ 0 & 0 & 1 \\ 1 & 1 & 1 \end{pmatrix}$ disjoint	$\begin{pmatrix} 0 & 0 & 1 \\ 0 & 1 & 1 \\ 1 & 1 & 1 \end{pmatrix}$ meet	$\begin{pmatrix} 1 & 0 & 0 \\ 0 & 1 & 0 \\ 0 & 0 & 1 \end{pmatrix}$ equal	$\begin{pmatrix} 1 & 1 & 1 \\ 1 & 1 & 1 \\ 1 & 1 & 1 \end{pmatrix}$ overlap
$\begin{pmatrix} 1 & 0 & 0 \\ 1 & 0 & 0 \\ 1 & 1 & 1 \end{pmatrix}$ inside	$\begin{pmatrix} 1 & 1 & 1 \\ 0 & 0 & 1 \\ 0 & 0 & 1 \end{pmatrix}$ contains	$\begin{pmatrix} 1 & 1 & 1 \\ 0 & 1 & 1 \\ 0 & 0 & 1 \end{pmatrix}$ covers	$\begin{pmatrix} 1 & 0 & 0 \\ 1 & 1 & 0 \\ 1 & 1 & 1 \end{pmatrix}$ coveredBy

The 9-intersection model grew from the earlier 4-intersection model, which did not include the exterior as part of the 4-intersection matrix [3, 11–13]. The 9-intersection model has since been extended to cover topological relations between multi-face regions (i.e., composite regions), regions with holes, complex regions, and to include information about the dimension of the of the interactions between objects [1, 2, 4, 9, 14, 16–19, 21, 22]. Topological relationships between map geometries were first studied in [15].

For reference, the complex types appear in Fig. 7.2. See [20] for a survey of spatial types in spatial database systems.

When considering topological relationships between map geometries, we are concerned with approaches to modeling topological relationships between regions. Despite the vast amount of research in topological relationships between various models of regions, the 8 topological relationships between simple regions form a useful foundation for characterizing many of the relationships among more complex models. Indeed, the 8 topological relationships between simple regions are straightforward, have intuitive names, and can be used to classify more complex relationships among more complex models. Figure 7.3 shows the 8 topological relationships between simple regions, their names, and their 9IMs.

The 9-intersection model of topological relationships is a *global* model in the sense that a single topological relationship exists for a given configuration of spatial objects. Such a global model can hide *local* information about the topological relationship between components of objects. The hiding of local information is expressed in two ways in global topological relationship models: through the *dominance problem*, and the *composition problem*. The dominance problem indicates the property that the global view exhibits *dominance* properties among the topological relationships as defined by the 9-intersection model. For example, while building roads between two adjacent countries, one might be interested to know that there is a disjoint island in one of the countries for which a bridge to the other country is required. The disjointedness in this case is overshadowed or dominated by the existing *meet* (adjacent) situation between the countries' mainlands. The composition problem expresses the property that a global topological predicate may indicate a certain relationship between two objects that does not exist locally. For example, consider two complex regions that have individual faces that satisfy the *inside*, *covers*, and *meet* predicates. Globally, this configuration satisfies the overlap predicate even though no faces overlap locally. These properties have been addressed through the *local topological relationship* models between composite regions [17] and between complex regions [18]. These approaches model a topological relationship between two multicomponent objects based on the topological relationships that exist between the components of the objects. We will use this idea to propose an implementation scheme for topological relationships between map geometries based on topological relationships between complex regions.

7.2 Data Model

We derive the topological relationships between the geometries imposed by spatial partitions. Therefore, the topological relationships between spatial partitions are independent of thematic values, and apply, in general, to partitions of the plane in which the regions are regular open point sets.

The 9-intersection model of topological relationships requires that spatial objects be defined based upon their interior, exterior, and boundary. We have provided such definitions for spatial partitions in Definition 2.1. The interior, exterior, and boundary of a spatial partition are illustrated in Fig. 7.4.

7.3 Topological Relationships Between Map Geometries

In this section, we describe a method for deriving the topological relationships between a given pair of map geometries. We begin by describing various approaches to the problem, then outline our chosen method, and finally derive the actual relationships based on this method.

The goal of this paper is to define a complete, finite set of topological predicates between map geometries; therefore, we employ a method similar to that found in [21], in which the 9-intersection model is extended to describe complex points, lines, and regions. Based on the model of spatial partitions, we extend the 9-intersection

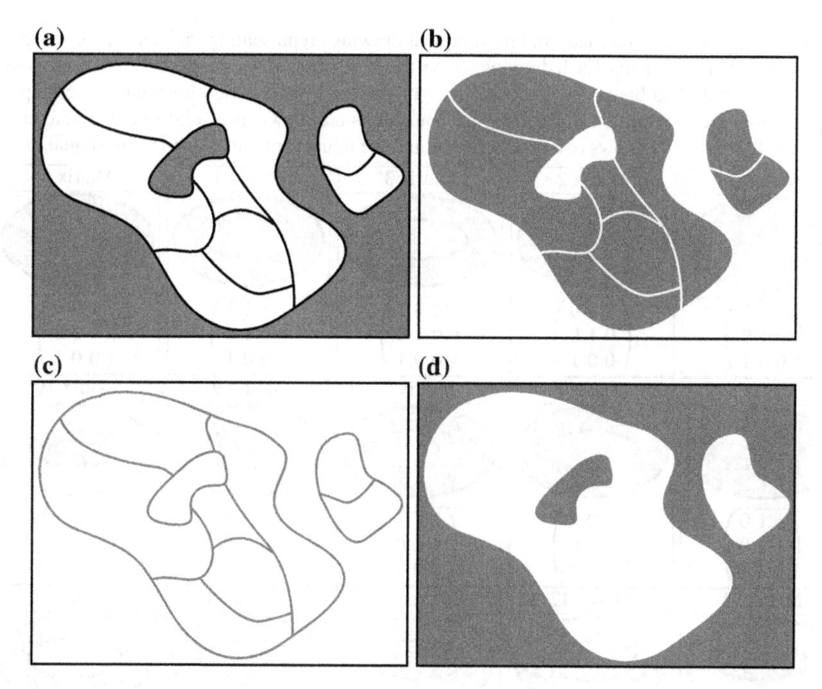

Fig. 7.4 **a** A spatial partition π with two disconnected faces, one containing a hole. **b** The interior (π°). **c** The boundary ($\partial\pi$). **d** The exterior (π^-). Note that the labels have been omitted in order to emphasize the components of the spatial partition

model to apply to the point sets belonging to map objects. However, due to the spatial features of map geometries, the embedding space ($\mathbb{R}^2$), and the interaction of map geometries with the embedding space, some topological configurations are impossible and must be excluded. Therefore, we must identify topological constraints that must be satisfied in order for a given topological configuration to be valid. Furthermore, we must identify these constraints such that all invalid topological configurations are excluded, and the complete set of valid configurations remains. We achieve this through a proof technique called *Proof-by-Constraint-and-Drawing*, in which we begin with the total set of 512 possible 9-intersection matrices, and determine the set of valid configurations by first providing a collection of topological constraint rules that invalidate impossible topological configurations, and second, validating all matrices that satisfy *all* constraint rules by providing a prototypical spatial configuration (i.e., the configurations can be drawn in the embedding space). Completeness is achieved because all topological configurations are either eliminated by constraint rules, or are proven to be possible through the drawing of a prototype. The remainder of this section contains the constraints, and the prototypical drawings of map geometries are shown in Table 7.1.

Table 7.1 The 49 valid matrices and prototypical drawings representing the possible topological relationships between maps. Each drawing shows the interaction of two maps, one map is medium-grey and has a dashed boundary, the other is light-grey and has a dotted boundary. Overlapping map interiors are dark-grey, and overlapping boundaries are drawn as a solid line. For reference, the figure for matrix 41 shows two disjoint maps and the figure for matrix 1 shows two equal maps

Matrix 1	Matrix 2	Matrix 3	Matrix 4	Matrix 5
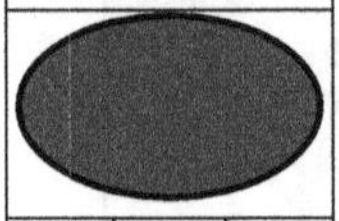	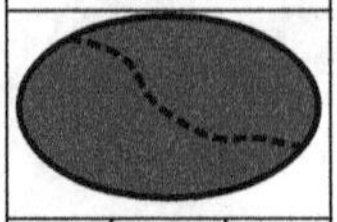	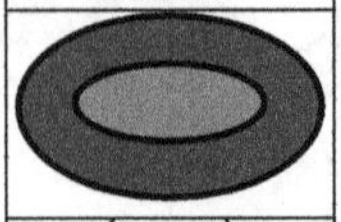	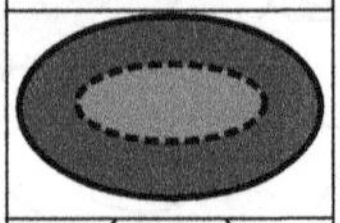	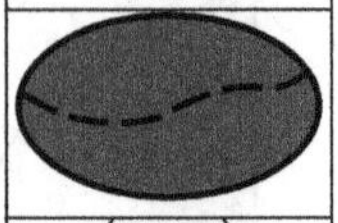
$\begin{pmatrix}1&0&0\\0&1&0\\0&0&1\end{pmatrix}$	$\begin{pmatrix}1&1&0\\0&1&0\\0&0&1\end{pmatrix}$	$\begin{pmatrix}1&0&1\\0&1&0\\0&0&1\end{pmatrix}$	$\begin{pmatrix}1&1&1\\0&1&0\\0&0&1\end{pmatrix}$	$\begin{pmatrix}1&0&0\\1&1&0\\0&0&1\end{pmatrix}$
Matrix 6	Matrix 7	Matrix 8	Matrix 9	Matrix 10
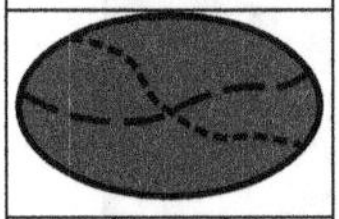	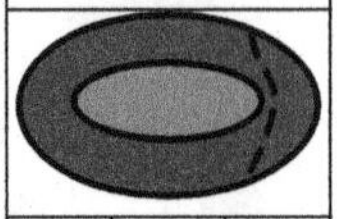	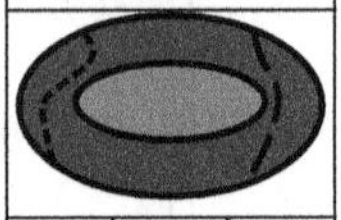	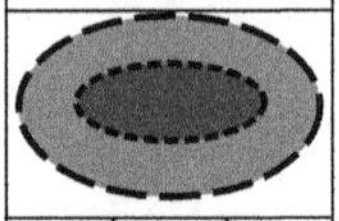	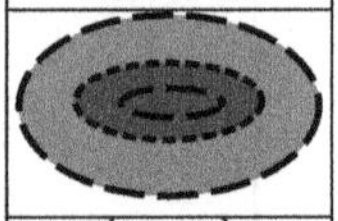
$\begin{pmatrix}1&1&0\\1&1&0\\0&0&1\end{pmatrix}$	$\begin{pmatrix}1&0&1\\1&1&0\\0&0&1\end{pmatrix}$	$\begin{pmatrix}1&1&1\\1&1&0\\0&0&1\end{pmatrix}$	$\begin{pmatrix}1&1&1\\0&0&1\\0&0&1\end{pmatrix}$	$\begin{pmatrix}1&1&1\\1&0&1\\0&0&1\end{pmatrix}$
Matrix 11	Matrix 12	Matrix 13	Matrix 14	Matrix 15
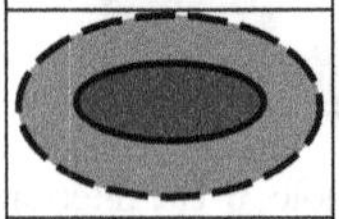	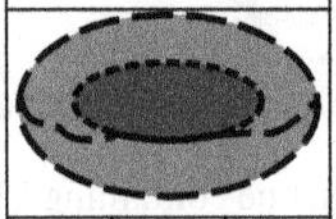	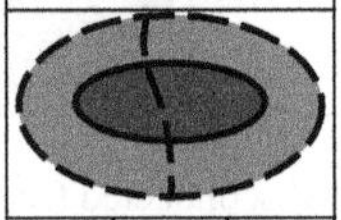	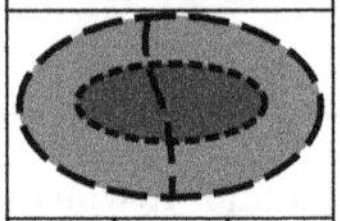	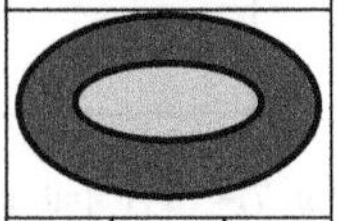
$\begin{pmatrix}1&0&1\\0&1&1\\0&0&1\end{pmatrix}$	$\begin{pmatrix}1&1&1\\0&1&1\\0&0&1\end{pmatrix}$	$\begin{pmatrix}1&0&1\\1&1&1\\0&0&1\end{pmatrix}$	$\begin{pmatrix}1&1&1\\1&1&1\\0&0&1\end{pmatrix}$	$\begin{pmatrix}1&0&0\\0&1&0\\1&0&1\end{pmatrix}$
Matrix 16	Matrix 17	Matrix 18	Matrix 19	Matrix 20
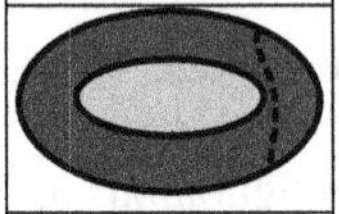		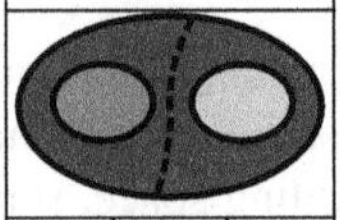	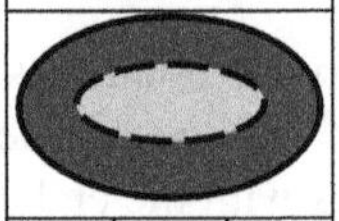	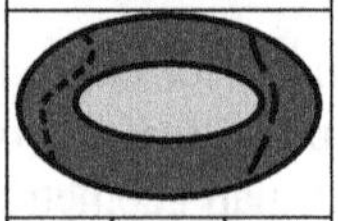
$\begin{pmatrix}1&1&0\\0&1&0\\1&0&1\end{pmatrix}$	$\begin{pmatrix}1&0&1\\0&1&0\\1&0&1\end{pmatrix}$	$\begin{pmatrix}1&1&1\\0&1&0\\1&0&1\end{pmatrix}$	$\begin{pmatrix}1&0&0\\1&1&0\\1&0&1\end{pmatrix}$	$\begin{pmatrix}1&1&0\\1&1&0\\1&0&1\end{pmatrix}$
Matrix 21	Matrix 22	Matrix 23	Matrix 24	Matrix 25
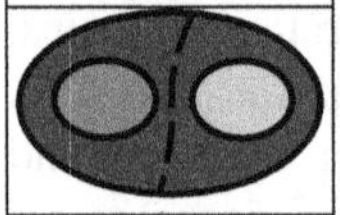		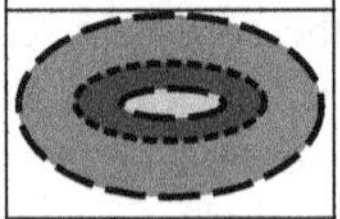	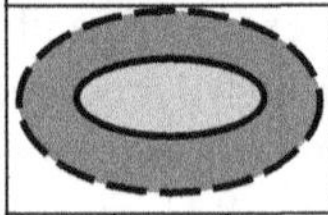	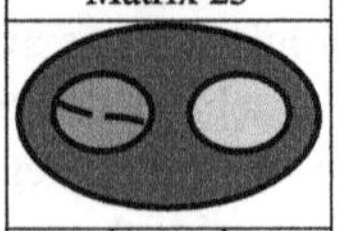
$\begin{pmatrix}1&0&1\\1&1&0\\1&0&1\end{pmatrix}$	$\begin{pmatrix}1&1&1\\1&1&0\\1&0&1\end{pmatrix}$	$\begin{pmatrix}1&1&1\\1&0&1\\1&0&1\end{pmatrix}$	$\begin{pmatrix}0&0&1\\0&1&1\\1&0&1\end{pmatrix}$	$\begin{pmatrix}1&0&1\\0&1&1\\1&0&1\end{pmatrix}$

(continued)

Table 7.1 (continued)

Matrix 26	Matrix 27	Matrix 28	Matrix 29	Matrix 30
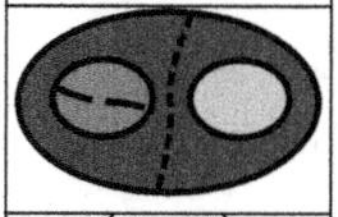	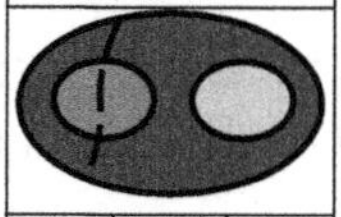	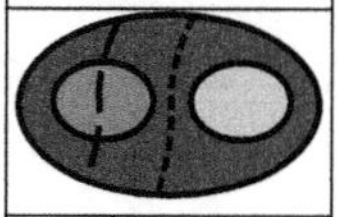	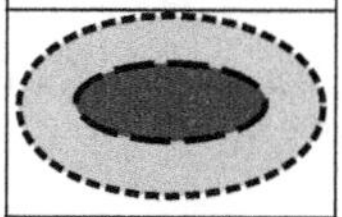	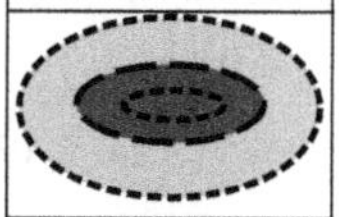
$\begin{pmatrix}1&1&1\\0&1&1\\1&0&1\end{pmatrix}$	$\begin{pmatrix}1&0&1\\1&1&1\\1&0&1\end{pmatrix}$	$\begin{pmatrix}1&1&1\\1&1&1\\1&0&1\end{pmatrix}$	$\begin{pmatrix}1&0&0\\1&0&0\\1&1&1\end{pmatrix}$	$\begin{pmatrix}1&1&0\\1&0&0\\1&1&1\end{pmatrix}$
Matrix 31	Matrix 32	Matrix 33	Matrix 34	Matrix 35
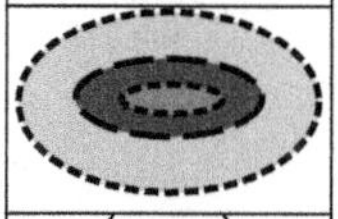	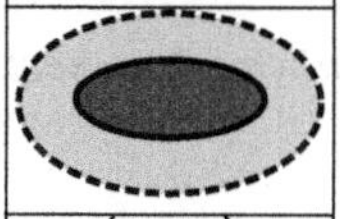	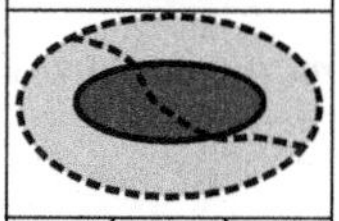	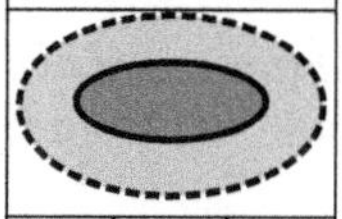	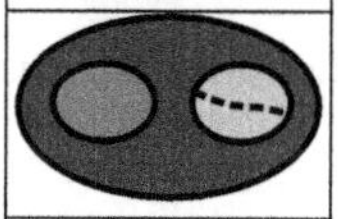
$\begin{pmatrix}1&1&1\\1&0&0\\1&1&1\end{pmatrix}$	$\begin{pmatrix}1&0&0\\0&1&0\\1&1&1\end{pmatrix}$	$\begin{pmatrix}1&1&0\\0&1&0\\1&1&1\end{pmatrix}$	$\begin{pmatrix}0&0&1\\0&1&0\\1&1&1\end{pmatrix}$	$\begin{pmatrix}1&0&1\\0&1&0\\1&1&1\end{pmatrix}$
Matrix 36	Matrix 37	Matrix 38	Matrix 39	Matrix 40
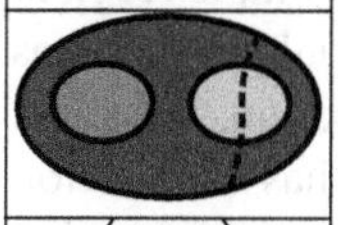	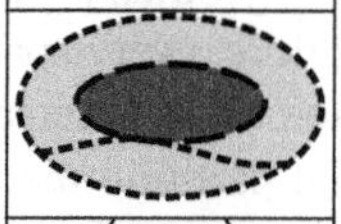	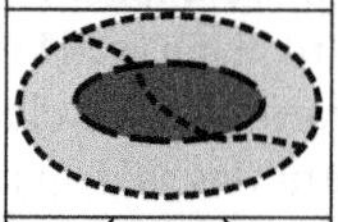	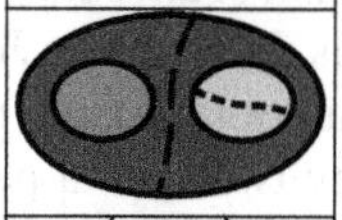	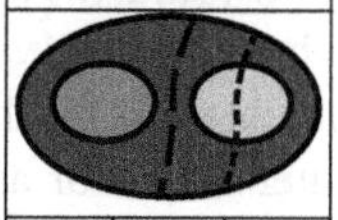
$\begin{pmatrix}1&1&1\\0&1&0\\1&1&1\end{pmatrix}$	$\begin{pmatrix}1&0&0\\1&1&0\\1&1&1\end{pmatrix}$	$\begin{pmatrix}1&1&0\\1&1&0\\1&1&1\end{pmatrix}$	$\begin{pmatrix}1&0&1\\1&1&0\\1&1&1\end{pmatrix}$	$\begin{pmatrix}1&1&1\\1&1&0\\1&1&1\end{pmatrix}$
Matrix 41	Matrix 42	Matrix 43	Matrix 44	Matrix 45
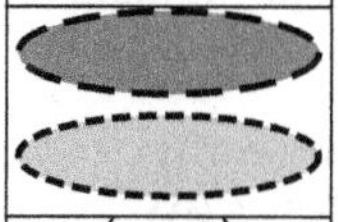	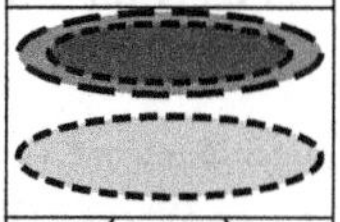	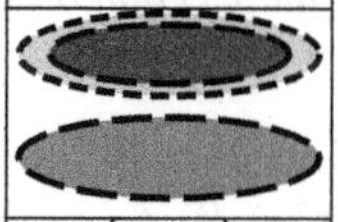	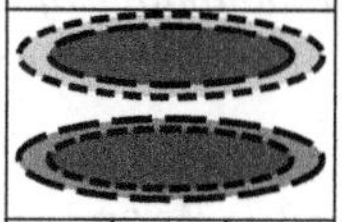	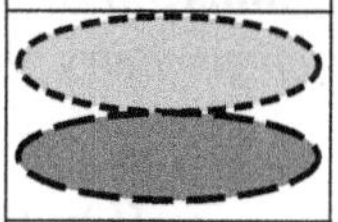
$\begin{pmatrix}0&0&1\\0&0&1\\1&1&1\end{pmatrix}$	$\begin{pmatrix}1&1&1\\0&0&1\\1&1&1\end{pmatrix}$	$\begin{pmatrix}1&0&1\\1&0&1\\1&1&1\end{pmatrix}$	$\begin{pmatrix}1&1&1\\1&0&1\\1&1&1\end{pmatrix}$	$\begin{pmatrix}0&0&1\\0&1&1\\1&1&1\end{pmatrix}$
Matrix 46	Matrix 47	Matrix 48	Matrix 49	
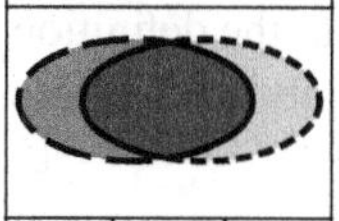	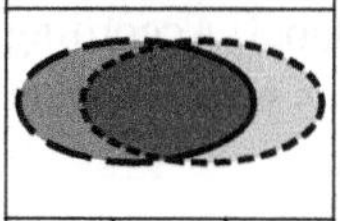	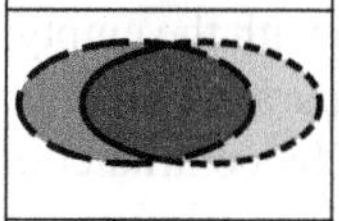	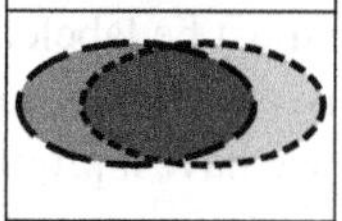	
$\begin{pmatrix}1&0&1\\0&1&1\\1&1&1\end{pmatrix}$	$\begin{pmatrix}1&1&1\\0&1&1\\1&1&1\end{pmatrix}$	$\begin{pmatrix}1&0&1\\1&1&1\\1&1&1\end{pmatrix}$	$\begin{pmatrix}1&1&1\\1&1&1\\1&1&1\end{pmatrix}$	

We identify eight constraint rules that 9IMs for map geometries must satisfy in order to be valid. Each constraint rule is first written in sentences and then expressed mathematically. Some mathematical expressions are written in two equivalent expressions so that they may be applied to the 9-intersection matrix more easily. Following each rule is the rationale explaining why the rule is correct. In the following, let π and σ be two spatial partitions.

Lemma 7.1 *Each component of a map geometry intersects at least one component of the other map geometry:*

$$(\forall C_\pi \in \{\pi^\circ, \partial\pi, \pi^-\} : C_\pi \cap \sigma^\circ \neq \emptyset \vee C_\pi \cap \partial\sigma \neq \emptyset \vee C_\pi \cap \sigma^- \neq \emptyset)$$
$$\wedge(\forall C_\sigma \in \{\sigma^\circ, \partial\sigma, \sigma^-\} : C_\sigma \cap \pi^\circ \neq \emptyset \vee C_\sigma \cap \partial\pi \neq \emptyset \vee C_\sigma \cap \pi^- \neq \emptyset)$$

Proof Because spatial mappings are defined as total functions, it follows that $\pi^\circ \cup \partial\pi \cup \pi^- = \mathbb{R}^2$ and that $\sigma^\circ \cup \partial\sigma \cup \sigma^- = \mathbb{R}^2$. Thus, each part of π must intersect at least one part of σ, and vice versa. □

Lemma 7.2 *The exteriors of two map geometries always intersect:*

$$\pi^- \cap \sigma^- \neq \emptyset$$

Proof The closure of each region in a map geometry corresponds to a complex region as defined in [21]. Since complex regions are closed under the union operation, it follows that the union of all regions that compose a map geometry is a complex region, whose boundary is defined by a Jordan curve. Therefore, every spatial partition has an exterior. Furthermore, spatial partitions are closed under intersection. Thus, the intersection of any two spatial partitions is a spatial partition that has an exterior. Therefore, the exteriors of any two spatial partitions intersect, since their intersection contains an exterior. □

Lemma 7.3 *If the boundary of a map geometry intersects the interior of another map geometry, then their interiors intersect:*

$$((\partial\pi \cap \sigma^\circ \neq \emptyset \Rightarrow \pi^\circ \cap \sigma^\circ \neq \emptyset) \wedge (\pi^\circ \cap \partial\sigma \neq \emptyset \Rightarrow \pi^\circ \cap \sigma^\circ \neq \emptyset))$$
$$\Leftrightarrow ((\partial\pi \cap \sigma^\circ = \emptyset \vee \pi^\circ \cap \sigma^\circ \neq \emptyset) \wedge (\pi^\circ \cap \partial\sigma = \emptyset \vee \pi^\circ \cap \sigma^\circ \neq \emptyset))$$

Proof Assume that a boundary b of partition π intersects the interior of partition σ but their interiors do not intersect. In order for this to be true, the label of the regions on either side of b must be labeled with the empty label. According to the definition of spatial partitions, a boundary separates two regions with different labels; thus, this is impossible and we have a proof by contradiction. □

Lemma 7.4 *If the boundary of a map geometry intersects the exterior of a second map geometry, then the interior of the first map geometry intersects the exterior of the second:*

$$((\partial\pi \cap \sigma^- \neq \emptyset \Rightarrow \pi^\circ \cap \sigma^- \neq \emptyset) \wedge (\pi^- \cap \partial\sigma \neq \emptyset \Rightarrow \pi^- \cap \sigma^\circ \neq \emptyset))$$
$$\Leftrightarrow ((\partial\pi \cap \sigma^- = \emptyset \vee \pi^\circ \cap \sigma^- \neq \emptyset) \wedge (\pi^- \cap \partial\sigma = \emptyset \vee \pi^- \cap \sigma^\circ \neq \emptyset))$$

Proof This proof is similar to the previous proof. Assume that the boundary b of partition π intersects the exterior of partition σ but the interior of π does not intersect the exterior of σ. In order for this to be true, the label of the regions on either side of b must be labeled with the empty label. According to the definition of spatial partitions, a boundary separates two regions with different labels; thus, this is impossible and we have a proof by contradiction. □

Lemma 7.5 *If the boundaries of two map geometries are equivalent, then their interiors intersect:*

$$
\begin{aligned}
&(\partial\pi = \partial\sigma \Rightarrow \pi^\circ \cap \sigma^\circ \neq \emptyset) \Leftrightarrow (c \Rightarrow d) \Leftrightarrow (\neg c \vee d) \text{ where}\\
&c = \partial\pi \cap \partial\sigma \neq \emptyset \wedge \pi^\circ \cap \partial\sigma = \emptyset \wedge \partial\pi \cap \sigma^\circ = \emptyset\\
&\qquad \wedge \partial\pi \cap \sigma^- = \emptyset \wedge \pi^- \cap \partial\sigma = \emptyset\\
&d = \pi^\circ \cap \sigma^\circ \neq \emptyset
\end{aligned}
$$

Proof Assume that two spatial partitions have an identical boundary, but their interiors do not intersect. The only configuration which can accommodate this situation is if one spatial partition's interior is equivalent to the exterior of the other spatial partition. However, according to Lemma 7.2, the exteriors of two partitions always intersect. If a partition's interior is equivalent to another partition's exterior, then their exteriors would not intersect. Therefore, this configuration is not possible, and the interiors of two partitions with equivalent boundaries must intersect. □

Lemma 7.6 *If the boundary of a map geometry is completely contained in the interior of a second map geometry, then the boundary and interior of the second map geometry must intersect the exterior of the first, and vice versa:*

$$
\begin{aligned}
&(\partial\pi \subset \sigma^\circ \Rightarrow \pi^- \cap \partial\sigma \neq \emptyset \wedge \pi^- \cap \sigma^\circ \neq \emptyset) \Leftrightarrow (\neg c \vee d) \text{ where}\\
&c = \partial\pi \cap \sigma^\circ \neq \emptyset \wedge \partial\pi \cap \partial\sigma = \emptyset \wedge \partial\pi \cap \sigma^- = \emptyset\\
&d = \pi^- \cap \partial\sigma \neq \emptyset \wedge \pi^- \cap \sigma^\circ \neq \emptyset
\end{aligned}
$$

Proof If the boundary of spatial partition π is completely contained in the interior of spatial partition σ, it follows from the Jordan Curve Theorem that the boundary of σ is completely contained in the exterior of π. By Lemma 7.4, it then follows that the interior of σ intersects the exterior of π. □

Lemma 7.7 *If the boundary of one map geometry is completely contained in the interior of a second map geometry, and the boundary of the second map geometry is completely contained in the exterior of the first, then the interior of the first map geometry cannot intersect the exterior of the second and the interior of the second map geometry must intersect the exterior of the first and vice versa:*

$$
\begin{aligned}
&((\partial\pi \subset \sigma^\circ \wedge \pi^- \supset \partial\sigma) \Rightarrow (\pi^\circ \cap \sigma^- = \emptyset \wedge \pi^- \cap \sigma^\circ \neq \emptyset))\\
&\Leftrightarrow (c \Rightarrow d) \Leftrightarrow (\neg c \vee d) \text{ where}\\
&c = \partial\pi \cap \sigma^\circ \neq \emptyset \wedge \partial\pi \cap \partial\sigma = \emptyset \wedge \partial\pi \cap \sigma^- = \emptyset\\
&\qquad \wedge \pi^\circ \cap \partial\sigma = \emptyset \wedge \pi^- \cap \partial\sigma \neq \emptyset\\
&d = \pi^\circ \cap \sigma^- = \emptyset \wedge \pi^- \cap \sigma^\circ \neq \emptyset
\end{aligned}
$$

Proof We construct this proof in two parts. According to Lemma 7.6, if $\partial\pi \subseteq \sigma^\circ$, then $\pi^- \cap \sigma^\circ \neq \emptyset$. Now we must prove that π° cannot intersect σ^-. Since $\partial\pi \subset \sigma^\circ$, it follows that π° intersects σ°. Therefore, the only configuration where $\pi^\circ \cap \sigma^- \neq \emptyset$ can occur is if σ contains a hole that is contained by π. However, in order for this configuration to exist, the $\partial\sigma$ would have to intersect the interior or the boundary of π. Since the lemma specifies the situation where $\pi^- \supset \partial\sigma$, this configuration cannot exist; thus, the interior of π cannot intersect the exterior of σ. □

Lemma 7.8 *If the boundary of a map geometry is completely contained in the exterior of a second map geometry and the boundary of the second map geometry is completely contained in the exterior of the first, then the interiors of the map geometries cannot intersect:*

$$((\partial\pi \subset \sigma^- \wedge \pi^- \supset \partial\sigma) \Rightarrow (\pi^\circ \cap \sigma^\circ = \emptyset))$$
$$\Leftrightarrow (c \Rightarrow d) \Leftrightarrow (\neg c \vee d) \text{ where}$$
$$c = \partial\pi \cap \sigma^\circ = \emptyset \wedge \partial\pi \cap \partial\sigma = \emptyset \wedge \partial\pi \cap \sigma^- \neq \emptyset$$
$$\wedge \pi^\circ \cap \partial\sigma = \emptyset \wedge \pi^- \cap \partial\sigma \neq \emptyset$$
$$d = \pi^\circ \cap \sigma^\circ = \emptyset$$

Proof The lemma states that the interiors of two disjoint maps do not intersect. Without loss of generality, consider two map geometries that each consist of a single region. We can consider these map geometries as complex region objects. If two complex regions are disjoint, then their interiors do not intersect. We can reduce any arbitrary map to a complex region by computing the spatial union of its regions. It follows that because the interiors of two disjoint regions do not intersect, the interiors of two disjoint maps do not intersect. □

Using a simple program to apply these eight constraint rules reduces the 512 possible matrices to 49 valid matrices that represent topological relationships between two maps geometries. The matrices and their validating prototypes are depicted in Table 7.1. Finally, we summarize our results as follows:

Theorem 7.1 *Based on the 9-intersection model for spatial partitions, 49 different topological relationships exist between two map geometries.*

Proof The argumentation is based on the *Proof-by-Constraint-and-Drawing* method. The constraint rules, whose correctness has been proven in Lemmas 7.1–7.8, reduce the 512 possible 9-intersection matrices to 49 matrices. The ability to draw prototypes of the corresponding 49 topological configurations in Table 7.1 proves that the constraint rules are complete. □

7.4 Computing Topological Relationships Between Map Geometries Using Topological Relationships Between Regions

Map geometries are composed of complex regions, and the topological predicates between complex regions are implemented in many spatial systems. Therefore, we now derive a method to determine the topological relationship between two map geometries based on the topological relationships between their component regions.

Recall that topological relationships based on the 9-intersection model are each defined by a unique 9IM. Given two map geometries, our approach is to derive the values in the 9IM representing their topological relationship by examining the *local* interactions between their component regions. We begin by discussing the properties of the 9IM, and then use these properties to show how we can derive a 9IM representing the topological relationship between two map geometries from the 9IMs representing the relationships between their component regions.

We begin by making two observations about the expression of local interactions between the component regions of map geometries in the 9IM that represents the geometries' relationship. The first observation is that if the interiors of two regions intersect, then the interiors of the map geometries intersect. This is because the presence of additional regions in a map geometry cannot cause two intersecting regions to no longer intersect. This observation also holds for interior/boundary intersections, and boundary/boundary intersections between regions composing map geometries. Therefore, if two map geometries contain respective regions such that those regions' interiors or boundaries intersect each other, then the 9IM for those regions will have the corresponding entry set to *true*.

The first observation does not hold for local interactions involving the exteriors of regions composing two map geometries. Consider Fig. 7.5. This figure depicts two map geometries, one with a solid boundary (which we name A), and a second one with a dashed boundary (which we name B) that is completely contained in the closure of the first. Note that if we examine the topological relationship between the leftmost regions in each geometry, it is clear that interior of the region from B intersects the exterior of the region from A. However, this does not occur globally due to the other regions contained in the geometries. However, we make the observation that the union of all regions that compose a map geometry is a complex region whose exterior is equivalent to the exterior of the map geometry. Therefore, a map geometry's interior or boundary intersects the exterior of a second map geometry if the interior or boundary of a single region in the first map geometry intersects the exterior of the union of all regions in the second. Therefore, we can discover the values in the 9IM representing the topological relationship between two map geometries by examining the interactions of those geometries' component regions and their unions. Given a map geometry π, we define the set of all complex regions in π as $R(\pi) = \{\overline{r} | r \in \rho(\pi)\}$ (recall that the regions in the set $\rho(\pi)$ are open point sets defining the interior of the regions; we take the closure to provide the boundary and interior of the complex region):

Definition 7.1 Let A and B be map geometries and $U_r = \bigcup_{r \in R(A)} r$ and $U_s = \bigcup_{s \in R(B)} s$. Let $M_{(A,B)}{}^{\circ\circ}$ be the entry in the matrix M representing a topological relationship between A and B corresponding to the intersection of the interiors of A

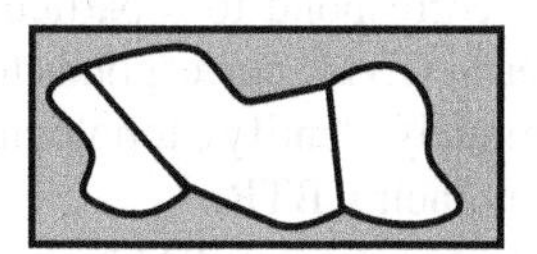 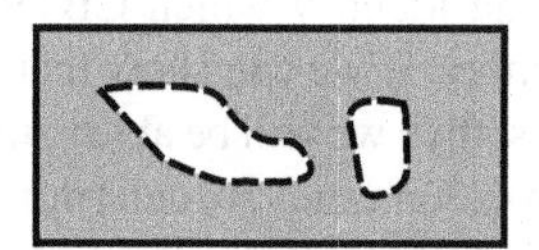 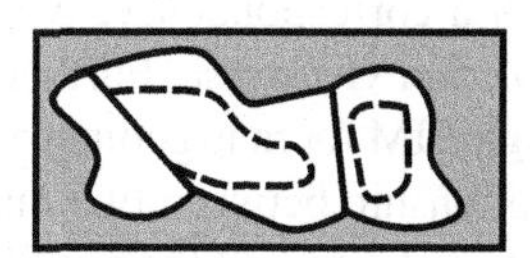

Fig. 7.5 Two example map geometries shown individually and overlaid

and B, etc. The entries in the 9IM representing the topological relationship between two map geometries can be defined as follows:

$$\begin{aligned}
M_{(A,B)}{}^{\circ\circ} &= \exists r \in R(A), s \in R(B) | r^{\circ} \cap s^{\circ} \neq \emptyset \\
M_{(A,B)}{}^{\circ}\partial &= \exists r \in R(A), s \in R(B) | r^{\circ} \cap \partial s \neq \emptyset \\
M_{(A,B)}\partial^{\circ} &= \exists r \in R(A), s \in R(B) | \partial r \cap s^{\circ} \neq \emptyset \\
M_{(A,B)}\partial\partial &= \exists r \in R(A), s \in R(B) | \partial r \cap \partial s \neq \emptyset \\
M_{(A,B)}{}^{-\circ} &= \exists s \in R(B) | U_r^- \cap s^{\circ} \neq \emptyset \\
M_{(A,B)}{}^{-}\partial &= \exists s \in R(B) | U_r^- \cap \partial s \neq \emptyset \\
M_{(A,B)}{}^{\circ -} &= \exists r \in R(A) | r^{\circ} \cap U_s^- \neq \emptyset \\
M_{(A,B)}\partial^{-} &= \exists r \in R(A) | \partial r \cap U_s^- \neq \emptyset
\end{aligned}$$

At this point we are able to characterize a 9IM representing the topological relationship between two map geometries based on intersections of the component regions that make up the map geometries. However, our goal is to go a step further and characterize such a relationship based on the topological predicates between complex regions that are currently implemented in spatial systems. In order to achieve this, we represent a topological relationship between map geometries as an ordered triple of sets of topological predicates between complex regions, which we denote a *component based topological relationship* (CBTR). This triple consists of a set of topological predicates between the component regions in two map geometries (from which we can directly identify whether the interiors, boundaries, or interiors and boundaries of the map geometries intersect), a set of topological predicates between the regions of the first map geometry and union of the regions in the second (to determine whether the interior and boundary of the first intersect the exterior of the second), and a set of topological predicates between the regions of the second map geometry and the union of the regions in the first.

Definition 7.2 Let P_{CR} be the set of topological predicates between complex regions. A component based topological relationship between map geometries A and B is an ordered triple $CBTR(A, B) = (P, N_1, N_2)$ of sets of topological predicates between complex regions such that:

$$\begin{aligned}
P &= \{p \in P_{CR} | r \in R(A) \wedge s \in R(B) \wedge p(r, s)\} \\
N_1 &= \{p \in P_{CR} | r \in R(A) \wedge U_s = \textstyle\bigcup_{s \in R(B)} s \wedge p(r, U_s)\} \\
N_2 &= \{p \in P_{CR} | U_r = \textstyle\bigcup_{r \in R(A)} r \wedge s \in R(B) \wedge p(U_r, s)\}
\end{aligned}$$

Definition 7.2 allows us to identify a CBTR given two map geometries. However, we still cannot determine the topological relationship between two map geometries given their CBTR. In order to do this, we must define a correspondence between 9IMs representing topological relationships between map geometries and CBTRs. Given such a correspondence, we can identify which CBTRs correspond to a particular 9IM, and vice versa. Furthermore, if we can show that each CBTR corresponds to a single 9IM for map geometries, then we will be able to uniquely identify a topological relationship between two map geometries by determining their CBTR.

In order to find the possible CBTRs that correspond to a particular 9IM between map geometries R, we find all possible values for each set in the triple (P, N_1, N_2).

We then find all combinations of these values that form a valid CBTR. To find all possible values of the set P, we take the powerset of the set of 9IMs representing topological relationships between complex regions. We then keep only the sets in the power set such that (i) for each interaction between interiors, boundaries, or interior and boundary in R with a value of *true*, at least one 9IM exists in the set that has a corresponding entry set to *true*, and (ii), for each interaction between interiors, boundaries, or interior and boundary in R with a value of *false*, all 9IMs in the set have a corresponding entry of *false*. This follows directly from the observations made about the intersections of interiors and boundaries among the regions that make up a map geometry. The set of possible values for N_1 and N_2 are computed identically, except entries corresponding to interactions involving the exterior of a map geometry are used. A CBTR that corresponds to the topological relationship R is then a triple (P, N_1, N_2) consisting of any combination of the computed values for each set. Because Definition 7.1 defines each entry in a 9IM based on an equivalence to information found in the CBTR, it follows that each CBTR corresponds to a single topological relationship between map geometries. Therefore, we are able to uniquely represent a topological relationship between map geometries as a CBTR, which consists of topological relationships between complex regions. To use this in a spatial database, one must compute the CBTR for two given map geometries, and then use the rules in Definition 7.1 to construct the 9IM that represents their topological relationship.

References

1. Clementini, E., Di Felice, P.: A comparison of methods for representing topological relationships. Information Sciences-Applications **3**(3), 149–178 (1995)
2. Clementini, E., Di Felice, P.: A model for representing topological relationships between complex geometric features in spatial databases. Information sciences **90**(1), 121–136 (1996)
3. Clementini, E., Di Felice, P., Van Oosterom, P.: A small set of formal topological relationships suitable for end-user interaction. In: Advances in Spatial Databases, pp. 277–295. Springer (1993)
4. Clementini, E., Di Felice, P., Califano, G.: Composite regions in topological queries. Information Systems **20**(7), 579–594 (1995)
5. Cohn, A.G.: Qualitative spatial representations. In: Proc. IJCAI-99 Workshop on Adaptative Spatial Representations of Dynamic Environments. Citeseer (1999)
6. Cohn, A.G., Hazarika, S.M.: Qualitative spatial representation and reasoning: An overview. Fundamenta informaticae **46**(1-2), 1–29 (2001)
7. Cohn, A.G., Bennett, B., Gooday, J., Gotts, N.M.: Qualitative spatial representation and reasoning with the region connection calculus. GeoInformatica **1**(3), 275–316 (1997)
8. Cui, Z., Cohn, A.G., Randell, D.A.: Qualitative and topological relationships in spatial databases. In: Advances in Spatial Databases, pp. 296–315. Springer (1993)
9. Egenhofer, M.J., Herring, J.: Categorizing binary topological relations between regions, lines, and points in geographic databases. Tech. rep., The National Center for Geographic Information and Analysis, University of California (1990)
10. Egenhofer, M.J., Herring, J.: A mathematical framework for the definition of topological relationships. In: Fourth international symposium on spatial data handling, pp. 803–813. Zurich, Switzerland (1990)

11. Egenhofer, M.J., Franzosa, R.D.: Point-set topological spatial relations. International Journal of Geographical Information System **5**(2), 161–174 (1991)
12. Egenhofer, M.J., Franzosa, R.D.: On the equivalence of topological relations. International Journal of Geographical Information Systems **9**(2), 133–152 (1995)
13. Egenhofer, M.J., Sharma, J., Mark, D.M.: A critical comparison of the 4-intersection and 9-intersection models for spatial relations: formal analysis. In: AUTOCARTO-CONFERENCE-, pp. 1–1. ASPRS AMERICAN SOCIETY FOR PHOTOGRAMMETRY AND (1993)
14. Egenhofer, M.J., Clementini, E., Di Felice, P.: Topological relations between regions with holes. International Journal of Geographical Information Science **8**(2), 129–142 (1994)
15. McKenney, M., Schneider, M.: Topological relationships between map geometries. In: Database Systems for Advanced Applications, pp. 110–125. Springer (2008)
16. McKenney, M., Pauly, A., Praing, R., Schneider, M.: Dimension-refined topological predicates. In: Proceedings of the 13th annual ACM international workshop on Geographic information systems, pp. 240–249. ACM (2005)
17. McKenney, M., Pauly, A., Praing, R., Schneider, M.: Preserving local topological relationships. In: Proceedings of the 14th annual ACM international symposium on Advances in geographic information systems, pp. 123–130. ACM (2006)
18. McKenney, M., Pauly, A., Praing, R., Schneider, M.: Local topological relationships for complex regions. In: Advances in Spatial and Temporal Databases, pp. 203–220. Springer (2007)
19. McKenney, M., Praing, R., Schneider, M.: Deriving topological relationships between simple regions with holes. In: Headway in Spatial Data Handling, pp. 521–531. Springer (2008)
20. Schneider, M.: Spatial Data Types for Database Systems - Finite Resolution Geometry for Geographic Information Systems, vol. LNCS 1288. Springer-Verlag, Berlin Heidelberg (1997)
21. Schneider, M., Behr, T.: Topological relationships between complex spatial objects. ACM Transactions on Database Systems (TODS) **31**(1), 39–81 (2006)
22. Worboys, M.F., Bofakos, P.: A canonical model for a class of areal spatial objects. In: Advances in Spatial Databases, pp. 36–52. Springer (1993)

Chapter 8
A Discrete Model of Maps

In previous chapters, we have defined an *abstract model* of maps called spatial partitions. We call that model an abstract model because it is defined based mathematical concepts, such as mappings of infinite point sets, that are not directly implementable in computer systems. The abstract model of spatial partitions maps each point in the plane to a specific label; However, computers provide only a finite resolution for the representation of data which is not adequate for the explicit representation of abstract spatial partitions. In order to represent maps in computers, a *discrete map model* is required that preserves the properties of spatial partitions while providing a representation that is suitable for storage and manipulation in computers. In this chapter, we provide a *graph model of spatial partitions*, which is a graph theoretic, discrete model of spatial partitions [2].

Although a discrete model of the data type of spatial partitions is designed around concepts that are discrete, it is distinct from an implementation model. For example, under the discrete model we assume that numbers are real numbers; in an implementation model, one must decide on a representation for real numbers, typically single or double precision floating point numbers. Thus, there are many possible implementation models corresponding to a discrete model, and many possible discrete models corresponding to an abstract model. The purpose of the discrete model is to preserve the properties of spatial partitions and enable implementation models.

There has been work in the literature that addresses the representation of maps as graphs. The work that comes closest to ours is [1, 3] in which the authors consider modeling maps as special types of plane graphs. However, the authors of these works define such graphs based on modeling a map as a collection type consisting of spatial point, line, and region objects. Problems in the proposed methods arise when different spatial objects in the map share coordinates. For example, given the method of deriving a plane graph from a collection of points, lines, and regions, it is unclear if a spatial point object that has the same coordinates as the endpoint of a spatial line object in the plane graph can be distinguished. Furthermore, the authors require a separate structure to model what they term the *combinatorial structure* of a plane graph, which includes the topological relationships between different spatial

M. McKenney and M. Schneider, *Map Framework*,
DOI 10.1007/978-3-319-46766-5_8

components of the graph. Finally, the plane graph, as defined, is not able to model thematic properties of the map.

8.1 Definitions from Graph Theory

There is some ambiguity among graph terms, especially concerning terms indicating graphs that are allowed to contain loops and multiple edges between pairs of vertices. We begin by first providing an overview of the graph terminology and definitions that we use to develop our model.

In graph theory, a graph is a pair $G = (V, E)$ of disjoint sets such that V is a set of vertices and $E \subseteq V \times V$ is a set of vertex pairs indicating edges between vertices. We denote the sets of vertices and edges for a given graph g as $V(g)$ and $E(g)$, respectively.

A *multigraph* is a pair $G_M = (V, E)$ of disjoint sets with a mapping $E \rightarrow V \times V$ allowing a multigraph to have multiple edges between a given pair of vertices.

A *loop* in a graph is an edge that has a single vertex as both of its endpoints. A multigraph with loops is a *pseudograph*, and is defined as a pair $G_P = (V, E)$ with a mapping $E \rightarrow V \times V$.

A *nodeless pseudograph* is a pseudograph that (possibly) contains edges that form loops that connect no vertices and that intersect no other edges or vertices. We define a nodeless pseudograph as a triple $G_N = (V, E, N)$ (where N is the set of nodeless edges) with mapping $E \rightarrow V \times V$.

A *path* is a non-empty graph $P = (V, E)$ such that $V = \{v_1, \ldots, v_n\}$, $E = \{v_1v_2, \ldots, v_{n-1}v_n\}$ and all v_i are distinct. Given a graph $g = (V, E)$, a path of g is a graph $p = (V_p, E_p)$ where $V_p \subseteq V(g) \wedge E_p \subseteq E(g)$ where p satisfies the definition of a path.

A *cycle* is a path whose first and last vertices are identical, defined by a non-empty graph $C = (V, E)$ such that $V = \{v_1, \ldots, v_n\}$, $E = \{v_1v_2, \ldots, v_nv_1\}$ where all v_i are distinct and $n \geq 3$. Given a graph $g = (V, E)$, a cycle of g is a graph $c = (V_c, E_c)$ where $V_c \subseteq V(g) \wedge E_c \subseteq E(G)$ where c satisfies the definition of a cycle. The *length* of a cycle is the number of its edges.

A *polygonal arc* is the union of finitely many straight line segments embedded into the plane and is homeomorphic to the closed unit interval $[0, 1]$. We use the terms *arc* and polygonal arc interchangeably.

A graph is *planar* if it can be embedded in the plane such that no two edges intersect. A particular drawing of a graph is an *embedding* of the graph. A particular planar graph can have multiple embeddings in the plane such that edges in each embedding are drawn differently. A particular embedding of a planar graph is a *plane graph*. Formally, a plane graph is a pair (V, E) such that $V \subseteq \mathbb{R}^2$, every edge is an arc between two vertices, the interior of an edge contains no vertex, and no two edges intersect except at their vertices. A plane graph g may contain cycles. We say a cycle c in plane graph g is *minimal* if there does not exist a path in g that splits the polygon induced by c into two pieces. We use the notation $C(g)$ to indicate the set of minimal cycles in the graph g.

8.2 Representing Spatial Partitions as Graphs

In this section, we define the type of *spatial partition graphs* that is able to model both the structural and labelling properties of spatial partitions in a graph. We first attempt to define a graph based on the vertex and edge structure of a partition, but show that this is not sufficient because the labels of the partition are not explicitly represented in such a graph. We then define a new type of graph that is capable of representing both the structural and labelling properties of a spatial partition and that is defined based on discrete concepts. We then show how such a graph can be obtained from a given spatial partition.

Recall that in the abstract model of spatial partitions (Chap. 2), we can identify a graph structure based on the boundary of a partition π. Specifically, we observe that we can identify $\nu(\pi)$, the set of points that are vertices, and $\varepsilon(\pi)$, the set of point sets belonging to edges of a partition. However, we cannot simply assign the vertices and edges to a pair (V, E) in order to achieve a graph representation of π since it is possible for nodeless edges to be present in $\varepsilon(\pi)$. For example, the boundary of the rightmost face of region A in Fig. 8.1a is composed of two nodeless edges; these edges are nodeless because the label of every point they contain is a set containing two attributes (i.e., no point in the edge satisfies the definition of a vertex in a partition). Therefore, we must use a triple (V, E, N) consisting of the sets of vertices, edges, and nodeless edges to represent a partition as a graph. Deriving the set V from a partition π is trivial because we can directly identify the set of vertex points $\nu(\pi)$. Definition 2.3 defines the set of all edges of π as $\varepsilon(\pi)$; thus, it does not differentiate between edges and nodeless edges. It follows that the set of nodeless edges of a partition, N, is a subset of $\varepsilon(\pi)$, but we cannot identify N by simply examining labels. Intuitively, the label of a nodeless edge should not form a subset of the label of any vertex. However, in Fig. 8.1b, the label of the boundary of the upper border of the left face of region A and the labels of both borders of the right face of region A are the same. Therefore, we cannot necessarily differentiate nodeless edges from edges by comparing labels. We can circumvent this problem if we can differentiate identically labeled edges from different faces of the same region. This can be achieved through the *refine* operation. Each nodeless edge in π can be identified as an edge in $\sigma = \mathit{refine}(\pi)$ whose label does not form a subset of any vertex label (a vertex lies on an edge in the refinement of a partition if the edge label is a subset of the vertex label). Note that the refine operation does not alter the edge structure of a partition, only its labels. Therefore, we can use $\sigma = \mathit{refine}(\pi)$ to identify the set of nodeless edges N in π by saying that each edge in σ whose label does not form a subset of any vertex label in σ is in the set N. The edges of π can then be calculated as $E = \varepsilon(\pi) - N$. Figure 8.1c shows the refinement of the partition in Fig. 8.1a. Figure 8.1d shows the graph representation of the partition

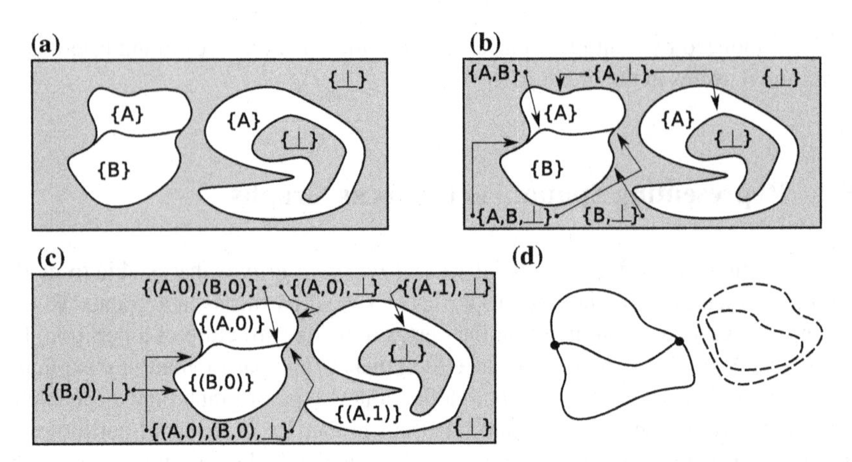

Fig. 8.1 A sample spatial partition with two regions. **a** The spatial partition annotated with region labels. **b** The spatial partition with its region and boundary labels. **c** The refinement of the partition in part (**a**). **d** The SSPG of the spatial partition. Nodeless edges are dashed

in Fig. 8.1a obtained by the method described above (nodeless edges are dashed). Because deriving a graph from a partition in this fashion results in a graph that exactly represents the edge structure of the partition from which it is derived, we call this type of graph a *structural spatial partition graph* (SSPG), and define it formally as follows:

Definition 8.1 Given a spatial partition π of type A and its refinement $\sigma = \mathit{refine}(\pi)$, we construct a structural spatial partition graph $SSPG = (V, E, N)$ with:

$$V = \nu(\pi)$$
$$E = \varepsilon(\sigma) - N$$
$$N = \{n \in \varepsilon(\sigma) | \nexists v \in \nu(\sigma) : \sigma(n) \subseteq \sigma(v)\}$$

The SSPG is able to represent the structural aspects of a spatial partition, but it does not maintain the labeling information of the partition. Because a SSPG is defined based on a given partition, the spatial mapping for the partition is known. Therefore, the label for any edge or vertex in a SSPG can be determined through the associated spatial mapping, but this is insufficient for our purposes as we require an explicit representation of labels. However, the SSPG has the property that it is a *planar nodeless pseudograph* (PNP):

Theorem 8.1 *Given a partition π of type A, its corresponding SSPG is a planar nodeless pseudograph.*

Proof The definition of a nodeless graph states that a nodeless graph may contain nodeless edges; therefore, an SSPG is nodeless by definition. Similarly, the definition of a pseudograph states that the graph may contain multiple edges between the same vertices and loops. An SSPG is therefore a pseudograph by definition because it does not exclude such features. Now we must prove that a SSPG is a planar graph. The edges of an SSPG are taken directly from a spatial partition, which is embedded in $\mathbb{R}^2$, indicating that the SSPG is an embedded graph. By the definition of spatial partitions, an edge in a partition is a border defined by a one-dimensional point set consisting of points mapped to a single label. Furthermore, this label is derived from the regions which the border separates. Assume that there exists two borders in a spatial partition that cross, which implies that the SSPG for this partition will contain two edges that intersect. In order for this to occur, these edges must separate regions that overlap, which violates the definition of spatial partitions. Therefore, a spatial partition cannot contain two borders that intersect, except at endpoints. Because the edges of an SSPG are taken directly from a spatial partition, then no two edges of an SSPG can intersect. Thus, the SSPG is a plane nodeless pseudograph. □

Although an SSPG can be easily obtained from a spatial partition, using a SSPG to model spatial partitions is inadequate for two reasons. First, spatial partitions depend on the concept of labels, so the graph representation of a partition must include a label representation. The SSPG does not implicitly model the labels of regions, edges, or vertices; rather, it depends on the existence of a spatial mapping. Second, the edges in the SSPG are taken directly from a spatial partition, which is defined on the concept of infinite point sets that we cannot directly represent discretely. Despite these drawbacks, the SSPG does have the nice property that because a SSPG is defined based on a given spatial partition, we know that any given SSPG is *valid* in the sense that it represents a spatial partition. Therefore, we proceed in two phases: we first define a type of graph that is capable of discretely representing the structural properties, and explicitly representing the labeling properties, of spatial partitions. We then show how we can derive graphs of this type from spatial partitions. This allows us to define a valid graph representation of any spatial partition.

It follows from Theorem 8.1 that an embedded graph that models a spatial partition such that its edges and vertices correspond to the partition's edges and vertices, respectively, must be a PNP. However, a PNP does not model the labeling of spatial partitions. In order to model labels in a graph, we must associate labels with some feature in a graph. In spatial partitions, the labels of boundaries can be derived from the labels of the regions they represent. Thus, it is possible to derive all edge and vertex labels in a partition from the region labels. Therefore, we choose to associate labels in a graph with features that are analogous to regions in spatial partitions. We are tempted to associate labels with minimum cycles in a graph representing a spatial partition; however, this is not able to accurately model situations in which a region in the spatial partition contains another region such that the boundaries of the regions are disjoint (e.g., the region labeled A_2 and the hole it contains in Fig. 8.1a). Instead, we associate labels with *minimum polycycles* (MPCs) in graphs representing partitions. A minimum polycycle in a PNP G is a set of minimum cycles consisting of a minimum cycle c_o (the outer cycle), and all other minimum cycles in G that lie within c_o, and not within any other minimum cycle. Note that a minimum cycle of a plane graph induces a region in the plane defined by a Jordan curve. Therefore, we can differentiate between the interior, boundary, and exterior of such a region. We denote the region induced in the plane by the minimum cycle C of plane graph G as $R(C)$. We now formally define minimum polycycles:

Definition 8.2 Let G be a planar nodeless pseudograph. A *minimum polycycle* of G is a graph $MPC = (V, E, N)$ where $V \subseteq V(G)$, $E \subseteq E(G)$, $N \subseteq N(G)$, and $C(MPC) \subseteq C(G)$, containing an *outer cycle* $c_o \in C(MPC)$ and zero or more *inner cycles* $c_1, \ldots, c_n \in C(MPC)$ such that:

(i) $\forall v \in V : (\exists d \in C(MPC) | v \in V(d))$
(ii) $\forall e \in E : (\exists d \in C(MPC) | e \in E(d))$
(iii) $\forall n \in N : (\exists d \in C(MPC) | n \in N(d))$
(iv) $\forall c_i \neq c_o \in C(MPC) : \partial R(c_i) \subseteq (\partial R(c_o) \cup R(c_o)^\circ)$
(v) $\nexists c_j, c_k \in C(MPC) | c_j \neq c_o \wedge c_k \neq c_o \wedge c_j \neq c_k$
$\wedge \partial R(c_j) \subseteq (\partial R(c_k) \cup R(c_k)^\circ)$
(vi) $\nexists d \in C(G) | (\partial R(d) \subseteq (\partial R(c_o) \cup R(c_o)^\circ)) \wedge \neg(d \in C(MPC))$
$\wedge (\forall c_i \neq c_o \in C(MPC) : \neg(\partial R(d) \subseteq (\partial R(c_i) \cup R(c_i)^\circ))$

In other words, an MPC is a sub-graph of a PNP G (Definition 8.2 parts $(i)-(iii)$) where the inner cycles are contained in the outer cycle (Definition 8.2 part (iv)), no inner cycle contains another inner cycle (Definition 8.2 part (v)) and there exists no cycle c in G that is not in the MPC such that c is contained in the outer cycle, but not contained in any inner cycles of the MPC (Definition 8.2 part (vi)).

Thus, a MPC induces a region in the plane that has a single face and that may contain holes defined by the minimum cycles that lie in the interior of the outer cycle. Furthermore, by associating labels with MPCs in a graph representing a spatial partition, we do not need to explicitly represent the labels of edges and vertices in the graph since they can be derived by simply finding all MPCs that an edge or vertex participates in. We denote the set of all MPCs for a PNP G as $MC(G)$.

The second problem with SSPGs is that the edges are defined as infinite point sets. We require a discrete representation of edges. Therefore, in addition to assigning labels to MPCs, we define our new type of graph such that its edges are arcs consisting of a finite number of straight line segments. We define the *labeled planar nodeless pseudograph* (LPNP) as a PNP with labeled MPCs, denoted *faces*, and edges modeled as arcs as follows:

Definition 8.3 Given an alphabet of labels Σ_L, a *labeled planar nodeless pseudograph* is defined by the four-tuple $LPNP = (V, E, N, F)$ consisting of a set of vertices, a set of arcs forming edges between vertices, a set of arc loops forming nodeless edges, and a set of faces, with:

$V \subseteq \mathbb{R}^2$
$E \subseteq V \times V \times (\mathbb{R}^2)^n$ where n is finite and each edge is an arc (arcs with endpoints in V and segment endpoints in $\mathbb{R}^2$)
$N \subseteq (\mathbb{R}^2)^n$ where n is finite (nodeless arc loops)
$F \subseteq \{(l \in \Sigma_L, m \in MC((V, E, N)))\}$

Recall that in the definition of spatial partitions, an unbounded face is explicitly represented with an empty label corresponding to the exterior of the partition. We do not explicitly model this unbounded face in the LPNP for two reasons: (i) we cannot guarantee that the edges incident to the unbounded face of a LPNP will form a connected graph, and (ii) if the edges incident to the unbounded face do form a connected graph, we cannot guarantee that it will be a cycle. However, we can determine if an edge in a LPNP is incident to the unbounded face if it participates in only a single MPC. This follows from the fact that each edge separates two regions in a partition. Because all bounded faces are modeled as MPCs in a LPNP, an edge that participates in only a single face must separate a MPC and the unbounded face.

The LPNP allows us to discretely model a labeled graph structure, but we have not yet discussed how we can obtain a LPNP for a given spatial partition. Note that because the edges of a LPNP are defined as arcs, they cannot directly represent edges from a partition. Instead, each edge in a LPNP is an approximation of an edge in its corresponding spatial partition. Therefore, we define the approximation function α that takes an edge from a partition and returns an arc which approximates that edge. α must return edges such that the vertices in the LPNP are identical to that of the corresponding spatial partition and the LPNP and the corresponding spatial partition are topologically identical. Given a spatial partition π of type A and an edge approximation function α, we derive the corresponding LPNP in a similar manner as we derived the SSPG from a partition. The set of vertices in the LPNP is equivalent to $\nu(\pi)$. In order to calculate the nodeless edges, we consider the refinement $\sigma = \mathit{refine}(\pi)$. Nodeless edges are then identified as all edges in σ whose label is not a subset of any vertex label. The set of edges is then difference of $\varepsilon(\pi)$ and the set of nodeless edges. Finally, each labeled MPC consists of the set of approximations of the edges surrounding each region in σ along with the label of the corresponding face in π. Given an edge e, we use the notation $V_e(e)$ to indicate the set of vertices that e connects. Figure 8.2 depicts the LPNP for the partition shown in Fig. 8.1.

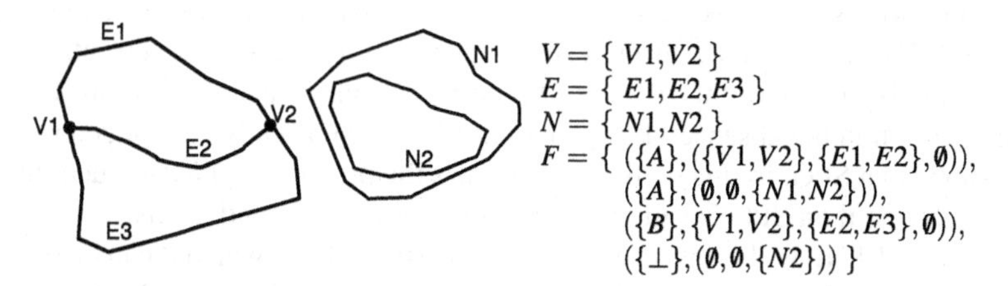

Fig. 8.2 A labeled plane nodeless pseudograph for the partition in Fig. 8.1. The edges and vertices are marked so that the sets of vertices, edges, nodeless edges, and faces can be expressed more easily

Definition 8.4 Given a spatial partition π of type A, edge approximation function α, and $\sigma = refine(\pi)$, we derive a $LPNP = (V, E, N, F)$ from π as follows:

$$
\begin{aligned}
V &= \nu(\pi) \\
E &= \{\alpha(e \in \varepsilon(\sigma) | \neg(\alpha(e) \in N))\} \\
N &= \{\alpha(n \in \varepsilon(\sigma) | \nexists v \in \nu(\sigma) : \sigma(n) \subseteq \sigma(v))\} \\
F &:= \forall r \in \rho(\sigma) | r \subseteq s \in \rho(\pi) : (l, (V_m, E_m, N_m)) \in F \; where : \\
&\quad l = \; \pi^{-1}[r] \\
&\quad E_m = \{\alpha(e) | e \in \omega(\sigma) \wedge e \subseteq \partial r \wedge \alpha(e) \in E\} \\
&\quad N_m = \{\alpha(n) | n \in \omega(\sigma) \wedge n \subseteq \partial r \wedge \alpha(n) \in N\} \\
&\quad V_m = \textstyle\bigcup_{e \in E_m} V_e(e)
\end{aligned}
$$

Note that it is possible to approximate edges in a partition in multiple ways. Thus, a single spatial partition may have multiple LPNPs that represent it.

In the definition of LPNPs, no restrictions are placed on the labels of MPCs. Therefore, it is possible for an labeled graph to fit the definition of an LPNP, but be labeled in such a way that violates the definition of spatial partitions. In other words, a LPNP may be labeled such that no spatial partition exists from which the LPNP can be derived. A simple example of this is a LPNP containing two MPCs that have the same label and share an edge. Because edges only separate regions with different labels in a partition, this LPNP can not be derived from any spatial partition. Thus, the set of all LPNPs is larger than the set of LPNPs that can be derived from some spatial partition. We define a LPNP that can be derived some spatial partition as a *spatial partition graph* (SPG).

Definition 8.5 A *spatial partition graph G* is a *labeled planar nodeless pseudograph* such that there exists some spatial partition from which G can be derived.

8.3 Properties of Spatial Partition Graphs

In the previous section, we defined the type of spatial partition graphs and showed how an SPG can be derived from a spatial partition. However, we currently define an SPG as being valid only if it can be derived from a spatial partition. Given a labeled graph in the absence of a spatial partition, we currently cannot determine if the graph is an SPG. In this section, we discuss the properties of SPGs such that we can determine if an SPG is valid by examining its structure and labels.

Given an LPNP in the absence of a source partition from which it can be derived, we must ensure that the structural and labeling properties of the LPNP are consistent with the properties of spatial partitions. To achieve this, we examine the properties of spatial partitions, defined by Definition 2.2 and Definition 2.3, and show how these properties are expressed in SPGs derived from spatial partitions. We then define the properties of SPGs and show that any LPNP that satisfies these properties is an SPG.

Recall that an SSPG is a PNP. It follows from Theorem 8.1 that any graph that models the edges of a spatial partition as graph edges and vertices of a spatial partition as graph vertices must be a PNP. Therefore, a graph cannot be an SPG if it is not a PNP. This is already expressed indirectly by the definition of SPGs as LPNPs.

Definition 2.2 formally defines constraints on spatial mappings that specify the type of spatial partitions. These constraints indicate that (i) the regions in a partition are regular open point sets, and (ii) the borders separate uniquely labeled regions and carry the labels of all adjacent regions. From these properties of partitions, we can derive properties of SPGs. By (i), we infer that all edges and vertices in a SPG must be part of an MPC. If an edge is not part of an MPC, then the edge does not separate two regions; instead, it extends either into the interior of an MPC or into the unbounded face of the SPG, forming a cut in the polygon induced by the MPC or the unbounded face. If a vertex exists that is not part of an MPC, then it is either a lone vertex with no edges emanating from it, or it is part of a sequence of edges that are not part of an MPC. In the first case, the vertex either exists within the region induced by an MPC or the unbounded face of the LPNP, forming a puncture. In the second case, the vertex exists in a sequence of edges that is not part of an MPC, which we have already determined to be invalid.

By the second property of spatial partitions (ii), we infer that edges must separate uniquely labeled MPCs. Therefore, there cannot be an edge in an LPNP that participates in two MPCs with the same label. Furthermore, every region in a partition must be labeled. It follows that every MPC in an SPG must be labeled. Because the unbounded face is not explicitly labeled in an SPG, one special case exists: a MPC forming a hole in a SPG (i.e., labeled with $\perp$) cannot share an edge with the unbounded face, as this would result in an edge separating two regions with the same label.

Definition 2.3 further identifies properties of spatial partitions. According to this definition, edges in spatial partitions always have two labels, and vertices always have three or more labels. Recall that in LPNPs, the unbounded face of the graph is not explicitly labeled. Therefore, in a SPG, all edges must participate in either one or two MPCs. Edges incident to the unbounded face of the graph will participate in only one MPC. The requirement that vertices have three or more labels in a spatial partition indicates that at a vertex, at least three regions meet. It follows that in a SPG, each vertex has at least a degree of three. Furthermore, because the unbounded face is not explicitly labeled, vertices must have at least two labels in a SPG (i.e., a vertex must participate in at least two MPCs). We summarize the properties of SPGs and show that any LPNP that satisfies these properties is a SPG:

Definition 8.6 An SPG G has the following properties:

(*i*) G is a planar nodeless pseudograph (Theorem 8.1)
(*ii*) $\forall e \in E(G) \cup N(G), \exists (l, X) \in F(G) | e \in E(X) \cup N(X)$ (Definition 2.2(i))
(*iii*) $\forall v \in V(G), \exists (l, X) \in F(G) | v \in V(X)$ (Definition 2.2(i))
(*iv*) $\forall v \in V(G) : degree(v) \geq 3$ (Definition 2.3)
(*v*) $\forall e \in E(G) \cup N(G) :$
$1 \leq |\{(l, X) \in F(G) | e \in E(X) \cup N(X)\}| \leq 2$ (Definition 2.3)
(*vi*) $\forall (l_1, X_1), (l_2, X_2) \in F(G) | l_1 = l_2 :$

$(\nexists e_1 \in E(X_1) \cup N(X_1), e_2 \in E(X_2) \cup N(X_2) | e_1 = e_2)$ (Definition 2.2(ii))

(vii) $\forall m \in MC(G), \exists f = (l, X) \in F(G) | m = X$ (Definition 2.2(ii))

$(viii)$ $\forall e \in E(G) \cup N(G) |$
$|\{(l, X) \in F(G) | e \in E(X) \cup N(X)\}| = 1 : l \neq \{\bot\}$ (Definition 2.2(ii))

Definition 8.6 indicates the following properties: (*i*) an SPG is a PNP. (*ii*) Each edge in an SPG is a member of some MPC. (*iii*) Every vertex is a member of some MPC. (*iv*) The degree if every vertex is 3 or greater. (*v*) Every edge in an SPG bounds exactly one or two MPCs. (*vi*) No two MPCs that are labeled identically share an edge. (*vii*) Every MPC in an SPG is a labeled face of the SPG (i.e., it has a label associated with it). (*viii*) For every edge that is contained by a single MPC, the label associate with that MPC cannot be the exterior label (an edge must bound exactly one or two areas containing the interior of the spatial partition).

Theorem 8.2 *Any LPNP that satisfies the properties in Definition 8.6 is a SPG.*

Proof The properties listed in Definition 8.6 indicate how the properties of spatial partitions are expressed in SPGs. From Theorem 8.1, we know that a valid SPG must be a PNP. Definition 2.2(i) states that all regions in a partition must be regular open sets. Because the faces in a SPG are analogous to regions in a spatial partition, this means that all edges and vertices must belong to some face; otherwise, they form a puncture or cut in some face of the SPG. Definition 8.6(ii) and Definition 8.6(iii) express this requirement. Definition 2.2(ii) states that borders in a partition between regions carry the labels of both regions. This implies that an edge in a spatial partition separates regions with different labels, and that every region in a spatial partition has a label. Definition 8.6(vi) and Definition 8.6(vii) express this by stating that if an edge participates in two faces of a SPG, those faces have different labels, and that every MPC in a SPG is a labeled face of the SPG. Because the unbounded face is not labeled, we must explicitly state that no edge that participates in a cycle forming a hole can have only a single label, as this implies that an edge is separating two regions with the $\{\bot\}$ label (Definition 8.6(viii)). Definition 2.3 states that edges in a partition carry two region labels, and that vertices carry three or more region labels. Because the unbounded face is not labeled in a SPG, we cannot directly impose these properties on a SPG. Instead, we observe that the number of region labels on a vertex in a partition indicates a minimum number of regions that meet at that vertex. Therefore, we can express this property in SPG terms by stating that vertices in a SPG must have degree of at least three (Definition 8.6(iv)). The edge constraint from Definition 2.3 can be specified in terms of a SPG by the property that an edge must participate in exactly one or two faces, indicating that the edge will have exactly one or two labels (Definition 8.6(v)). Therefore, all properties of partitions are expressed in terms of SPGs in Definition 8.6, and any LPNP that satisfies these properties is a SPG. □

8.4 Deriving Spatial Partitions from Spatial Partition Graphs

We now have the ability to either derive a valid SPG from a spatial partition, or verify that a SPG is valid in the absence of a spatial partition from which it can be derived. Finally, we show that given an SPG, we can directly construct a spatial partition that exactly models the SPG's spatial structure and labels. Recall that a spatial partition is defined by a spatial mapping that maps points to labels and satisfies certain properties. Therefore, to construct a partition from a SPG, we must be able to derive a spatial mapping from a SPG.

We construct a spatial mapping based on the labeled MCPs of a SPG. Each MCP of a SPG induces a polygon in the plane that is associated with a label. Each of these polygons is a spatial region, and is defined by its boundary, which separates the interior of the polygon from its exterior. Therefore, we can identify the interior, boundary, and exterior of such a polygon. We use the notation $R(X)$ to denote the polygon induced in the plane by MPC X. A point that falls into the interior of a polygon can therefore be mapped directly to that polygon's label.

The labels of points belonging to edges in the SPG are slightly more difficult to handle. Each point belonging to an edge is mapped to the labels of each face in which the edge participates. If an edge happens to be incident to the unbounded face (it is an edge participating in a single MPC), it also is mapped to the $\{\perp\}$ label. Similarly, each vertex is mapped to the labels of each cycle in which it participates. If a vertex is incident to the unbounded face, it is also mapped to the $\{\perp\}$ label. A vertex is incident to the unbounded face if and only if it is the endpoint of an edge that is incident to the unbounded face. In order to define this mapping, we first provide a notation to distinguish between edges and vertices that are incident to the unbounded face, and those that are not. We then show how to derive a spatial partition from a SPG:

Definition 8.7 Let G be an SPG. We distinguish two sets of edges: the set containing edges incident to the unbounded face, denoted $E_\perp$, and the set containing edges not incident to the unbounded face, denoted E_b. Likewise, we distinguish two sets of vertices: the set containing vertices incident to the unbounded face, denoted $V_\perp$, and the set containing vertices not incident to the unbounded face, denoted V_b. These sets are defined as follows:

$$
\begin{aligned}
E_\perp(G) &= \{e \in E(G) \cup N(G) \mid |\{(l, X) \in F(G) | e \in E(X) \cup N(X)\}| = 1\} \\
E_b(G) &= (E(G) \cup N(G)) - E_\perp(G) \\
V_\perp(G) &= \{v \in V(G) | (\exists e \in E_\perp | v \in V_e(e))\} \\
V_b(G) &= V(G) - V_\perp(G)
\end{aligned}
$$

A spatial partition is constructed from an SPG as follows:

Definition 8.8 Let G be an SPG. We can directly construct a spatial partition π of type A as follows:

$$A = \{l | (l, X) \in F(G)\} \cup \{\perp\}$$

$$\pi(p) = \begin{cases} \{l | (l, X) \in F(G) \wedge p \in R(X)^\circ\} & \text{if } \exists (l, X) \in F(G) | p \in R(X)^\circ \quad (1) \\ \{\perp\} & \text{if } \nexists (l, X) \in F(G) | \\ & \quad p \in R(X)^\circ \cup \partial R(X) \quad (2) \\ \{l | (l, X) \in F(G) \wedge p \in \partial R(X)\} & \text{if } (\exists (l, X) \in F(G) | p \in \partial R(X)) \\ & \quad \wedge (\nexists e \in E_\perp(G) | p \in e) \quad (3) \\ \{l | (l, X) \in F(G) \wedge p \in \partial R(X)\} \cup \{\perp\} & \text{if } (\exists (l, X) \in F(G) | p \in \partial R(X)) \\ & \quad \wedge (\exists e \in E_\perp(G) | p \in e) \quad (4) \end{cases}$$

In Definition 8.8, the type of a spatial partition derived from a SPG consists of the set of labels of faces in the SPG plus the exterior label. The spatial mapping is then defined by finding the labels of all faces a point participates in. If a point p lies in the interior of a face (1), then the label of that point will be the label of the face. If p does not lie in the interior or boundary of any face (2), it maps to the label $\{\perp\}$. If p lies on a boundary that is not incident to the unbounded face (3), it is mapped to the set of labels of all faces that include p in its boundary. Note that because face boundaries include vertex points, this case handles the mapping of both edge and vertex points. Similarly, if p lies on a face boundary that is incident to the unbounded face (4), then p is mapped to the set of labels from all faces which include p, as well as the unbounded face label.

8.5 Summary

In this chapter, we have shown a discrete model of spatial partitions that utilizes graphs whose edges are polygonal arcs. Polygonal arcs are straightforward to implement in computer systems since they are constructed of straight line segments and a straight line segment is represented simply as a pair of end points. Furthermore, we have shown that a spatial partition graph represents a spatial partition, and that a spatial partition can be directly constructed from a spatial partition graph; therefore, type closure of operations defined on spatial partitions translate to spatial partition graphs. The final step in developing spatial partitions is to create an implementation model of spatial partitions suitable for implementation in computer systems.

References

1. De Floriani, L., Marzano, P., Puppo, E.: Spatial queries and data models. In: Spatial Information Theory A Theoretical Basis for GIS, pp. 113–138. Springer (1993)
2. McKenney, M., Schneider, M.: Spatial partition graphs: A graph theoretic model of maps. In: Advances in Spatial and Temporal Databases, pp. 167–184. Springer (2007)
3. Viaña, R., Magillo, P., Puppo, E., Ramos, P.A.: Multi-vmap: A multi-scale model for vector maps. Geoinformatica **10**(3), 359–394 (2006)

Chapter 9
Implementing Maps: Map2D

At this point, we have an abstract model of the Map Framework that precisely defines the semantics of maps and their operations and predicates. Furthermore, we have shown how to represent maps in a discrete manner such that they maintain the properties of the abstract model. In this chapter, we provide an implementation model of the Map Framework, denoted *Map2D*, consisting of a data model for maps and algorithms for computing map operations [18]. Furthermore, we show how operations over maps can be combined to form new map operations.

9.1 Data Model

We propose a data model with the goal of implementing maps in database management systems (DBMSs); therefore, we describe a data model in the context of an SQL database. This does not decrease generality since the data model simply assumes tables consisting of tuples; indeed, this can be implemented in a variety of ways.

The data model for spatial partitions in a database system is composed of two components: the geometric component and the label component. The geometric component consists of the map geometry, and is implemented as a column type named *map2D*. This is similar in concept to the implementation of traditional spatial types in databases [14, 27, 30, 32, 33]. The difference between map2D and the implementation of traditional types is that each region in the map is associated with a region identifier that will be used to associate each region with its attribute data.

The second component to the map2D type consists of the attributes for each region. We assume that each instance of a map2D column is associated with a *label table*, which contains the attributes for each region in the map. The only constraint we place on the label table is that it must contain a column of type integer named *region_id* that is used to associate each tuple in the label table with a region in its

M. McKenney and M. Schneider, *Map Framework*,
DOI 10.1007/978-3-319-46766-5_9

MapTable

ID: int	Name: str	Geom: map2D
1	Map_A	1 2
2	Map_B	1 2

Map1AttributeTable

region_id: int	Crop: str
1	Wheat
2	Corn

Map2AttributeTable

region_id: int	Avg Temp: int	Avg Rain: int
1	98	14
2	86	22

Fig. 9.1 A relation containing a map2D column and the associated label tables. The table Map1AttributeTable is associated with the map with ID equal to 1 in the table MapTable, and the table Map2AttributeTable is associated with the map with ID equal to 2 in the table MapTable

corresponding map. By storing the region attributes in a separate database table, instead of within the map2D data type itself, we allow the user to perform queries over the attributes associated with regions using standard SQL. Finally, we assume that a map instance stores the information necessary to identify its corresponding label table. Figure 9.1 shows an example instance of a table containing a column of type map2D which contains two tuples, and the label tables associated with the map stored in each tuple.

It may be more practical to have a single label table for all maps in a column, and introduce a map identifier column to the label table; however, we will assume a separate label table for each map for clarity of examples.

The geometry portion of a map is more complex than the label portion. Our goal is to incorporate the geometry portion of a map directly as a column type in a database. Therefore, we must implement the geometry portion as a single object that can be incorporated into a DBMS through extension mechanisms or through extending the DBMS code itself. There are several considerations that must be addressed in the design of the geometry: (i) the map geometry must support map operations that will be performed over it; and (ii) the concept of regions must be maintained in the geometry such that map regions must be able to be identified, associated with labels in a label table, and reconstructed.

Recall that that nearly all map operations are expressed using the the power operations, *intersect*, *relabel*, and *refine* (Chap. 3) [9, 10, 25]. Refine does not affect the geometric structure of a map, and the relabel operation has minimal geometric impact, at most regions are merged. The intersect operation, however, requires two maps to be merged both geometrically and thematically. The geometric merging can be computationally intensive; thus, we design the geometric data model to support algorithms to compute the intersection of maps. The discrete data model of spatial partition graphs defines the geometric portion of maps to be implemented as collections of straight line segments that represent the map borders. Thus, the intersection of two maps requires line segment intersection algorithms [2–5, 12, 13, 24, 26, 31, 34]. The line segment intersection algorithm with optimal time complexity [3, 5] is the *plane sweep algorithm*, so we design the geometric portion of the map2D type with the plane sweep algorithm in mind.

9.1.1 Overview of the Plane Sweep Algorithm

The optimal time complexity to compute the intersections between a set of n line segments in the plane using the plane sweep algorithm is $O(n \lg n + k)$ where k is the number of line segment intersections discovered [3, 5]; thus, the plane sweep algorithm is *output sensitive* since its time complexity depends on the number of line segment intersections that exist in the input set of line segments. In the worst case, nearly every line segment will intersect nearly every other line segment, leading to a value of k approaching n^2, but such a configuration is rare in practice, especially in geographic contexts. We will begin our discussion of the plane sweep algorithm by considering the *overlay* of two regions. Recall that a region is a special case of a spatial partition in which the the type of the spatial partition contains only two elements, the external region symbol, and a symbol that identifies a single region. The overlay operation is identical to the intersect operation on spatial partitions. Therefore, the overlay operation must manage both geometric and label aspects of the input regions. Furthermore, the overlay is concerned with finding the intersections of pairs of segments such that each segment in the pair belongs to distinct input regions; this is referred to as a *red-blue* intersection problem [6, 17, 29].

Let R and S be two regions. The plane sweep algorithm proceeds by sweeping an imaginary line, denoted the *sweep line* over R and S. Each time the line encounters a new line segment on the boundary of one of the input geometries, that segment is added to a list of segments actively being considered, known as the *active list*. Each time the line moves past a segment, that segment is removed from the active list. The active list stores the line segments in the order in which they intersect the sweep line; thus, if the sweep line is traveling in the x direction of a euclidean plane, the segments in the active list are sorted based on the y value of their intersection point with the imaginary sweep line.

Because segments in the active list are sorted, the segments in the active list adjacent to a segment s from region R that is newly inserted into the active list contains the information necessary to determine whether s lies on the interior, boundary, or exterior of W. This knowledge is deduced because each line segment carries two identifiers, one indicating the region that lies above the segment, and the other indicating the region that lies below the segment. If no region lies above or below a segment, then a default label indicating the exterior of the region is assigned. The plane sweep algorithm *processes* one segment in each iteration of the algorithm. Processing a segment involves adding it to the active list, discovering if it intersects any other line segments by examining its immediate neighbors in the active list, and determining which region interiors lie above and below the segment by looking at its immediate neighbors in the active list.

9.1.2 Implementation Aspects of the Plane Sweep Algorithm

In implementation, the sweep line does not move continuously in a sweep line algorithm, but instead progresses based on segment endpoints. Therefore, each segment is actually represented twice, once for its beginning end point, and once for its ending end point, and the segments are sorted. Because segments are represented twice, the notion of a *halfsegment* is used to represent them. The definition of halfsegments is based on segments. A segment is a straight line segment between two end points.

$$\begin{aligned} segment = (p, q) | p = (p_x, p_y) \wedge q = (q_x, q_y) \wedge p_x, p_y, q_x, q_y \in \mathbb{R} \\ ((p_x < q_x) \vee (p_x = q_x \wedge p_y < q_y)) \end{aligned} \tag{9.1}$$

Let [*segment*] indicate the type of all segments defined as the set of all possible segments. A halfsegment is a tuple containing a segment, a boolean value, and two *labels* respectively corresponding to the portion of the embedding space that lies above the segment (or to the left in the case of a vertical segment), and the portion of the embedding space that lies below the segment (or the right in the case of a vertical segment). We assume labels are integers that may be used as identifiers corresponding to thematic values stored elsewhere.

$$halfsegment = \{(s, b, l_a, l_b) | s \in [segment], b \in \mathbb{B}, l, r \in \mathbb{Z}\} \tag{9.2}$$

Let [*halfsegment*] indicate the type of halfsegments defined as the set of all possible halfsegments.

Thus, a halfsegment is said to have two *sides*, a left side and right side, corresponding to each label. As a matter of notation, we refer to l_a as the *above label* of halfsegment h, and l_b as the *below label* of h. For a halfsegment $h = (s, b, l_a, l_b)$, if b is true (false), the smaller (greater) endpoint of s is the *dominating point* of h, and h is called a *left* (*right*) *halfsegment*. Hence, each segment s is mapped to two halfsegments $(s, True, l_a, l_b)$ and $(s, False, l_a, l_b)$.

Let *dp* be a function which yields the dominating point of a halfsegment:

$$\begin{aligned} &dp : [halfsegment] \rightarrow \mathbb{R}^2 \\ &dp(h) := \begin{cases} p | h = ((p, q), b, l_a, l_b) & \textbf{if } b \\ q | h = ((p, q), b, l_a, l_b) & \textbf{otherwise} \end{cases} \end{aligned} \tag{9.3}$$

Let *len* be a function that returns the length of a halfsegment:

$$
\begin{aligned}
&len : [halfsegment] \rightarrow \mathbb{R} \\
&len(h) := \sqrt{(q_x - p_x)^2 + (q_y - p_y)^2} \ |h = ((p, q), b, l_a, l_b)
\end{aligned} \tag{9.4}
$$

Let *collinear* be a function that returns a value of *True* if two line segments are collinear.

$$
\begin{aligned}
&collinear : [segment] \times [segment] \rightarrow \mathbb{B} \\
&collinear(s_1 = (p, q), s_2 = (u, v)) := \\
&\quad \begin{cases} True & \textbf{if } ((v_y - u_y) * (q_x - p_x)) - ((v_x - u_x) * (q_y - p_y)) = 0 \\ False & \textbf{otherwise} \end{cases}
\end{aligned} \tag{9.5}
$$

The *rotate* predicate will take two halfsegments h_1 and h_2. *rotate* returns a value of *True* if a ray r that extends vertically from the shared end point of h_1 and h_2 intersects h_1 before it intersects h_2 when r is rotated in a counter-clockwise direction around the shared end point. Otherwise, *rotate* returns *False*. A halfsegment that extends vertically from the shared end point is not considered to intersect r until r has rotated 360°.

$$
rotate : [halfsegment] \times [halfsegment] \rightarrow \mathbb{B} \tag{9.6}
$$

A total ordering exists over points. let p and q be points:

$$
p < q \Leftrightarrow (p_x < q_x \vee (p_x = q_x \wedge p_y < q_y)) \tag{9.7}
$$

The plane sweep algorithm requires a total ordering over halfsegments:

$$
\begin{aligned}
&\text{Let } h_1 \in [halfsegment] | h1 = (s_1, b_1, l_{a1}, l_{b1}) \\
&\text{Let } h_2 \in [halfsegment] | h2 = (s_2, b_2, l_{a2}, l_{b2}) \\
&h_1 < h_2 \Leftrightarrow \\
&dp(h_1) < dp(h_2) \vee \\
&(dp(h_1) = dp(h_2) \wedge (\\
&\quad (\neg b_1 \wedge b_2) \vee \\
&\quad (b_1 = b_2 \wedge rotate(h_1, h_2)) \vee \\
&\quad (b_1 = b_2 \wedge collinear(s_1, s_2) \wedge len(s_1) < len(s_2)) \\
&\quad) \\
&)
\end{aligned} \tag{9.8}
$$

Figure 9.2 depicts an example sequence of the sweep line algorithm. In Fig. 9.2, part of the sweep line algorithm is shown in which two triangles are overlaid. Line segments are shown with their labels on each side. The sweep line is dotted, and the segments currently in the active list are shown dashed. The sweep line visits halfsegments in halfsegment order. The labels of all halfsegments behind the sweep line are finalized, anything in front of the sweep line must still be processed. Note

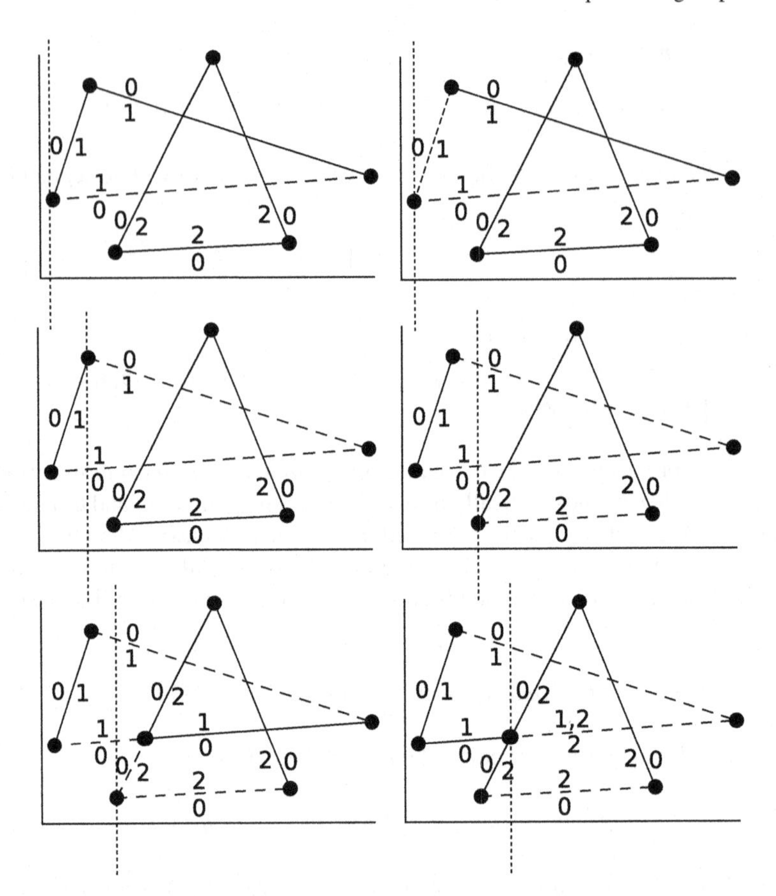

Fig. 9.2 A partial example sequence of a sweep line algorithm overlaying two triangles. Segments are shown with labels indicating where triangle interiors lie. Note that only labels to the left of the sweep line are finalized, as labels to the right of the sweep line have not yet been processed. The label 0 indicates the exterior of both triangles

the final step depicted: when the halfsegment is added to the sweep line, it is clear (based on the labels of the segment being added and the labels of the segment below it in the sweep line) that the interiors of both triangles intersect above the segment. This intersection of triangle interiors is indicated in the labels of the segment. The illustration shows the label above that segment containing two integers, which is not allowed by our definition of halfsegments; instead, a new integer identifier must be created that is associated with the integers indicating the interiors of all regions that lie above that segment. Therefore, the sweep line progression provides the ability to compute the necessary geometric information to integrate regions into maps. Furthermore, the opportunity to integrate thematic information exists, but the complexity of handling thematic information is non-trivial since an arbitrary number of regions may intersect. We address a mechanism to associate identifiers with labels indicating that multiple regions overlap in later sections.

9.2 Building Maps

9.2.1 Preparing the Input

The first problem to address in implementing the map2D type for the Map Framework is the problem of building an instance of the map2D type [20, 21]. As indicated above, a map2D instance will be a collection of halfsegments forming the geometric portion of the map such that labels on halfsegments correspond to attribute values in an associated label table. We must first be able to build a collection of halfsegments that forms a map2D geometry. Let $S \subseteq [segment]$ be a set of line segments; A method to construct a set of halfsegments that represent the region induced in the plane by R is described in [19]. That method uses a plane sweep algorithm, has time complexity of $O(n \lg n)$ where $n = |S|$, and will construct labels such that the interior of the the region will have a single, user defined, integer label, and the exterior will also have a single, user defined, integer label. Thus, we begin the construction of a map with a set of regions R where each region in the set is defined as a set of appropriately labeled halfsegments such that the interior label for each region in R is unique. Finally, we must sort all the halfsegments in all regions into a single list in halfsegment order. Let $m = |R|$ be the number of regions in R and c be the number of halfsegments in all m regions; preparing the input takes $O(c \lg c)$ time since all halfsegments must be sorted (assuming a comparison sort).

9.2.2 Spatial Processing of a Halfsegment

Once the list of sorted halfsegments for all regions that must be integrated into a map is computed, the plane sweep portion of the algorithm commences. We assume a sweep line that travels from left to right across the euclidean plane in the x direction. The plane sweep portion of the algorithm serves two main functions: (i) detect and remedy line segments that intersect at points other than end points, and (ii) merge the labels of the intersecting portions of regions. These functions are managed simultaneously within the algorithm.

Recall that the plane sweep algorithm proceeds by processing a single halfsegment at a time. At each step of the algorithm in which a left halfsegment h is being processed, h is inserted into the active list, and the immediate neighbors of h in the active list are computed; these neighbors are a and b, the neighbor halfsegment that lies above h and below h in the active list, respectively. The intersection points between h and a, and h and b are computed, the segment $s \in \{a, b\}$ that forms the least intersection point p with h in halfsegment order is chosen. h and s are removed from the active list, split according to p, and the resulting segments that are less than h in halfsegment order are inserted back into the active list. This step is the traditional computation of line segment intersections performed by the sweep line.

9.2.3 Thematic Processing of a Halfsegment

Once the spatial portion of the processing of halfsegment h is complete, the second function of the algorithm proceeds. The goal of the second part of the algorithm is to maintain thematic information associated with regions in the presence of multiple overlapping regions. Recall that if regions overlap in the spatial partition map model, the overlapping portions of the regions carry the labels of all the overlapping regions.

We proceed with this discussion using examples. Consider Fig. 9.3a in which two regions overlap. The regions have identifiers, which we assume are reflected in the halfsegment's labels. The segments in the figure are numbered. For figures where segments are numbered, we use the notation 1_l and 1_r to indicate the left and right halfsegments corresponding to segment 1, respectively. The plane sweep begins and processes halfsegments 1_l, 2_l, 2_r, and 3_l, all of which belong to the same region and do not intersect any other halfsegments. When the plane sweep algorithm begins to process 5_l, the active list contains 1_l and 3_l; 1_l is the neighbor below 5_l and 3_l is the neighbor above 5_l in the active list. It follows from halfsegment ordering that the above label of the halfsegment below 5_l in the active list will indicate which region 5_l lies inside, or if it lies in the exterior of all other regions. Therefore, 5_l lies in the interior of region 1 because the interior of region 1 lies above 1_l, and this will be reflected in its above label. Because the interior of region 2 lies above 5_l, it follows that the interiors of both region 1 and region 2 lie above 5_l. From the spatial partition definition, we follow the convention that overlapping portions of regions receive the labels of all regions involved in the overlap; therefore a new label is assigned to the overlapping portion, and the labels of 5_l must be changed to reflect the new topology (label management details will be discussed below).

When the plane sweep algorithm processes 4_l in Fig. 9.3, it will only have one neighbor in the active list: 5_l. An intersection is detected between the two, and the segments are split. Because no segments exist in the active list below 4_l, it cannot lie in the interior of another region, and its labels remain unchanged. If the overlapping portion of region 1 and region 2 receives the identifier 3, then Fig. 9.4b indicates the result of the algorithm. Because the label 3 indicates the overlapping portion of regions 1 and 2, some record keeping must exist to indicate that region identifier 3 is equivalent to the pair of region identifiers 1 and 2.

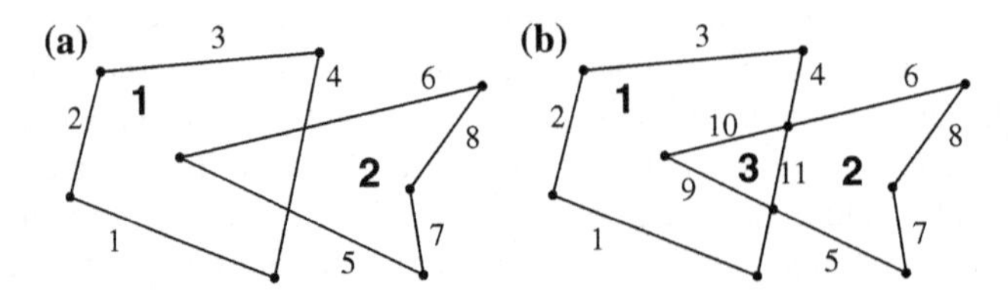

Fig. 9.3 Two regions superimposed (**a**), and the result of their integration into a map (**b**). Segments are numbered

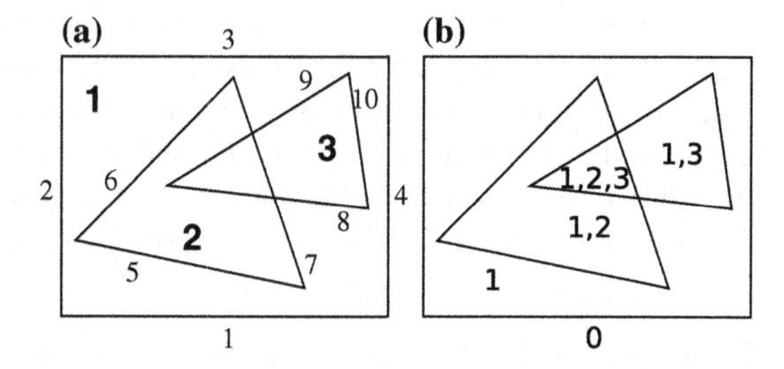

Fig. 9.4 Three overlapping regions. In **a**, segments are numbered and the original region labels are shown in bold. **b** shows the region labels indicating which region interiors lie in each region primitive. *0* indicates the exterior of all regions

Managing the labels of a pair of overlapping regions, as in the example above, is relatively straightforward; however, difficulty arises in cases where multiple regions overlap. For example, in Fig. 9.4 three regions are shown, all of which overlap other regions. Again, computing the line segment intersections is straightforward, but managing region identifiers becomes much more challenging. For instance, when segment 8_l is processed, the halfsegment below it in the active list is 5_l. In order for us to correctly label the above label of 8_l as the overlap of regions 1, 2, and 3, the above label of 5_l must indicate that its above label is the overlap of regions 1 and 2, or we must proceed further down the active list than the halfsegment immediately below the one currently being processed. The time complexity bounds of the plane sweep algorithm rely on the fact that only the immediate neighbors of a segment in the active list are examined when processing the current segment, so we cannot look beyond the immediate neighbors without causing an $O(n^2)$ time complexity. Therefore, we must devise a mechanism such that when a halfsegment h is processed, all regions whose interiors intersect h can be identified by examining only the halfsegment below h in the active list. In essence, the labels of h must be merged with the labels of the halfsegment below h in the active list to reflect the overlapping regions; we denote this *label merging*.

A successful label merge of a halfsegment results in the halfsegment's labels indicating the identifiers of all regions that lie immediately above and below the halfsegment. Therefore, the label of a halfsegment must be a set of region identifiers indicating all regions whose interiors lie on either side of the halfsegment. For example, when halfsegment 8_l in Fig. 9.4a is processed, it must indicate that the interiors of regions 1 and 2 lie immediately below the halfsegment, and the interiors of regions 1, 2, and 3 lie immediately above it. The original above and below labels of halfsegment 8_l are 3 and 0, respectively, where 0 indicates the exterior of the region. When 8_l is processed, we can determine the identifiers of all regions that must be involved in labels of 8_l by looking at the above label of the halfsegment below it in the active list, 5_l. Because 5_l is less than 8_l in halfsegment order, it will be processed before 8_l and its labels will be known when 8_l is processed. The above label of 5_l must indicate that the interiors of regions 1 and 2 lie above it. Therefore, the identifiers for regions 1 and 2 must be merged into 8_l's labels, indicating that 8_l lies within regions 1 and 2.

When 9_l is processed, the halfsegment below it in the active list, 8_l, indicates that region identifiers 1, 2, and 3 must be merged with the labels of 9_l. Because the labels of 9_l indicate that it bounds region 3, the identifier for region 3 will not be added to the above label of 9_l (i.e., the interior of region 3 does not extend above 9_l). Therefore, the above and below labels of 9_l must indicate the region identifiers 1 and 2, and 1, 2, and 3, respectively. Formally, a label merge is represented as the following:

Definition 9.1 Let $h = (p, q, A, B)$ be a halfsegment and $h_b = (p_b, q_b, A_b, B_b)$ be the halfsegment below h in the active list when h is being processed by the plane sweep algorithm. Because h is not processed, it will have the identifier of the input region it bounds as one of its labels, and the other will be the identifier of the exterior. A *label merge* is then:
$A = A \cup (A_b - B)$
$B = B \cup (A_b)$
These formulas follow directly from the definition of regions and the fact that a boundary always separates the interior of a region from the exterior of a region.

A single exception to the above definition exists when a halfsegment is being processed, and a spatially equivalent halfsegment (i.e., the halfsegments have identical endpoints and are both left or right halfsegments) is already present in the active list. Such a situation is especially common when integrating maps consisting of political divisions, which typically contain many regions that are adjacent and share boundary segments. Because the new halfsegment does not lie completely above the spatially equivalent halfsegment in the active list, it does not lie in the interior of the region primitive bounded by the halfsegment in the active list. Therefore, the new halfsegment cannot obtain the labels of all region interiors that lie below the halfsegment by examining only the labels of regions that lie above the halfsegment in the active list. Instead, the labels of all region interiors that lie *above* the new halfsegment can be determined from the labels of region interiors that lie above the corresponding halfsegment in the active list, and the region interiors that lie *below* the halfsegment must be determined by the region interiors that lie below the corresponding halfsegment in the active list. Using the original label merge technique will cause a halfsegment s to appear to be lying on the interior of regions who's interiors do not actually include s. Formally, a label merge for two spatially identical halfsegments is defined as follows.

Definition 9.2 Let $h = (p, q, A, B)$ be a halfsegment and $h_b = (p_b, q_b, A_b, B_b)$ be a halfsegment with identical end points to h that is in the active list when h is being processed by the plane sweep algorithm. Because h is not processed, it will have the identifier of the input region it bounds as one of its labels, and the other will be the identifier of the exterior. A *label merge* for spatially identical halfsegments is then:
$A = A \cup (A_b - B)$
$B = B \cup (B_b)$
These formulas follow directly from the definition of regions and the fact that a boundary always separates the interior of a region from the exterior of a region.

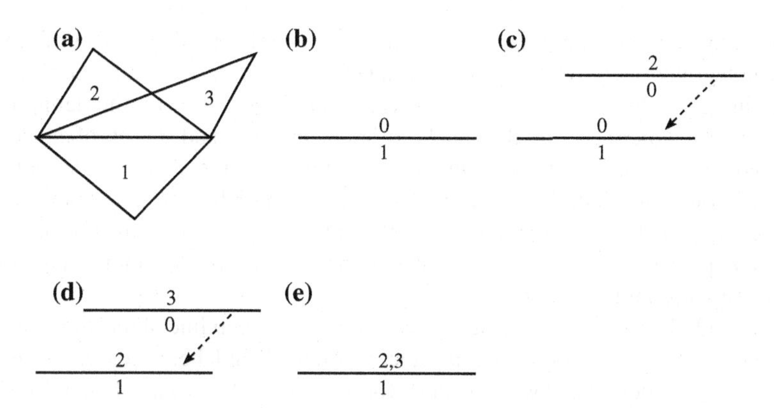

Fig. 9.5 A possible sequence of encountering spatially identical segments with different labels in the plane sweep portion of the algorithm. **a** shows the three triangular regions that are to be integrated into a map. The remainder of the figures show the sequence in which spatially identical halfsegments are integrated into a map, and the resulting labeling. By using the label merge formula for spatially identical halfsegments, the correct labeling will be computed

Figure 9.5 depicts an example of merging region labels over three spatially identical segments, each from a different region. If the standard label merge is used, the label for region 1 will be incorrectly propagated above the segment shown in Fig. 9.5b when the segment from region 2 is added. As the sequence of figures indicates, using the label merge for spatially identical halfsegments ensures the correct labeling scheme propagates through the resulting map.

When a halfsegment is being processed, it will always have exactly one identifier in one of its labels, and the exterior identifier in the other label. The halfsegment below it in the active list may have many region identifiers in its above and below labels. It follows from the definition of halfsegment merging that computing the below label of the current halfsegment h involves removing a single region identifier from the above label of the halfsegment below h in the active list. Furthermore, removing the below label of h from the above label of the halfsegment below h in the active list and inserting the above label of h also involves removing and adding single region identifiers. Therefore, the label merges can be expressed as three separate operations in which a single region identifier is added to, or subtracted from, a set of region identifiers. When integrating m regions, the size of the set of region identifiers on the halfsegment below the halfsegment being processed is bounded by m; therefore, efficient insertion and removal operations are required. We assume these sets of region identifiers are implemented as a height bounded binary search tree that supports removal and insertion in $O(\lg m)$ time.

If many regions overlap in a map, then a label may contain many region identifiers. Storing a list of region identifiers for each segment forming the boundary of a map is inefficient in terms of space, especially when a map must be stored on disk; therefore, we maintain a mapping that defines a relationship between *label identifiers* and labels. A label identifier is a unique identifier that corresponds to a label. Because a label may contain many region identifiers, it is bounded by the number of regions being integrated (m^2 for m regions). A label identifier can be implemented as a single integer, which is much more storage efficient and fits with the definition of halfsegments which specifies that labels are single integers, and not sets of integers. A hash table is a natural choice for implementing this mapping, and provides constant time insertions and lookups. A reverse mapping from labels to label identifiers is also required and can be implemented efficiently using a trie, providing insertion and lookup in $O(m)$ time for m input regions [7, 11], or as another hash table, providing constant time operations.

When a halfsegment h is being processed by the sweep line algorithm, the halfsegment below h is computed and its label is found. That label is copied twice, and a label merge is performed with the label for the region above h and below h respectively with each copy. The result of the label merges are labels indicating all regions whose interiors lie immediately above and below h. Because many halfsegments may contain identical labels, and these labels may contain up to m region identifiers for m input regions, representing each label with a label identifier can drastically reduce the memory requirements of the algorithm in situations when labels contain many region identifiers. Therefore, a trie associating labels with label identifiers is queried to see if the computed label is already assigned to another halfsegment and already has an associated label identifier. If the label has not yet been encountered, then a new label identifier is created for the label, and the trie is updated. A hash table is also updated to reflect the mapping from the label identifier to its corresponding label. Therefore, for each halfsegment visited by the plane sweep algorithm, a trie must be queried and possibly modified twice, and a hash table must be queried and possibly modified twice, leading to a time complexity of $O(m + c)$ for m regions for each halfsegment visited by the plane sweep (c is a constant). The optimal plane sweep has a time complexity of $O(n \lg n + k)$ for n segments with k intersections [3, 5]; however, the version of plane sweep that is used more typically in practice has complexity $O(n \lg n + k \lg n) \Leftrightarrow O((n + k) \lg n)$ where $n + k$ segments are visited and the $\lg n$ term reflects operations on data structures necessary for each segment visited by the algorithm. Therefore, the overall time complexity of the algorithm is $O((n + k)(\lg n + m + c))$.

The space complexity of the algorithm depends on the number of unique labels created during the map building. In the worst case, every region will overlap every other region in such a way that m^2 labels will exist for m regions. Because each unique label is stored, and every segment from regions must be stored, the space complexity is $O(m^2 + n + k)$ for m regions containing n line segments with k intersections. In practice, the number of unique labels generated is typically much less then m^2; furthermore, m is typically much smaller than n, especially in geographic data sets, so a larger space or time complexity on the m term is acceptable.

9.2.4 A Note on External Memory

The usage of the plane sweep algorithm in an external memory setting has been considered in the realm of databases where the plane sweep algorithm can be used to perform spatial joins over relations containing regions [1, 16]. Because the plane sweep algorithm reads its input from a sorted list of halfsegments, it is clear that the entire list need not be kept in memory. Therefore, one may implement a buffered disk reader that reads a chunk of halfsegments into memory at a time for the plane sweep to take as input. Organizing the plane sweep active list as an external memory structure is less clear; however, the size of the plane sweep active list tends to follow the *square root rule* [15, 28], meaning the size of the active list does not, in practice, exceed the square root of the number of input halfsegments. Because of this observation, creating an external memory active list is not necessary; experimental results have confirmed this [21]. Furthermore the labeling structures are typically small since, especially in geographic data sets; the number of intersecting regions typically is a linear function of the number of input regions, rather than the quadratic worst case scenario.

Finally, the resulting map that is output from the algorithm must be written to disk as it is produced. However, the halfsegments produced by the plane sweep are not necessarily produced in the correct order. The version of the plane sweep algorithm that does produce its output in sorted order is, in practice, complex and difficult to implement [3, 5]. The original plane sweep design is slightly less efficient, but significantly more straightforward to implement, and does not produce a sorted output; so, a sorting step is required for the output. An external memory sort algorithm is sufficient.

9.2.5 A Note on Topological Predicates

Topological predicates are trivially computed by the map building algorithm. The result of the map building algorithm provides the information necessary to determine which region interiors from the input regions intersect. The only modification is that one must record which region boundaries intersect when computing line segment intersections, since the intersection of two region interiors does not necessarily imply a boundary intersection. This process can be used to directly compute topological predicates between maps, or to compute topological predicates between component regions of maps, and then use the method presented in Chap. 7 to compute topological predicates between maps using the topological predicates between their component regions.

9.3 The Intersection Operator

The intersection operator takes two maps as input, and overlays them such that any point in the embedding space is contained a region carrying the labels of all regions that contain that point in either of the input maps. Figure 9.6 depicts an example.

A region is a special case of a map in which the map contains only a single region. The map construction algorithm takes a set of regions and constructs a map from them; however, there is nothing in its definition that prohibits using maps as input. Care must be taken that input maps do not share label identifiers on halfsegments, but no other restrictions exist. Therefore, the geometric portion of the intersect operator simply uses the map construction algorithm. We assume, based on the definition of the intersect operator, that intersect is a binary operator.

The final portion of the intersect operator requires that a new label table be constructed based on the information from the label tables in the two input maps. The result label table contains all columns from the label table of the original maps, except it will contain only a single *region_id* column. Furthermore, each column name will have the integer 1 or 2 appended if it came from the label table in the first or second argument map to the intersect operation, respectively. This ensures that no naming conflicts occur.

Finally, we must populate the result label table. First, all labels for regions in the first input map that have portions that did not intersect any region from the second map are added to the label table. These regions can be determined during the plane sweep and kept in a list. Second, the same procedure is done for regions in the second input map that contain a portion that does not overlap any region from the first input map. Note that for these labels, the values in columns from the opposing map will be given default or empty values. Then, the mapping from merged region labels to new region labels is used to construct the labels for overlapping regions from the input maps.

For example, assume that the two regions in Fig. 9.3a each belong to a separate map and those maps are used as input to an intersect operation. Table 9.1 shows a label table for each region, and the label table for the result of the intersect operator applied to those maps. The algorithm to compute the label table is shown in Algorithm 1.

The lists of labels that are in the label table resulting from an intersection of two maps are constructed during the plane sweep portion of the algorithm, as is the

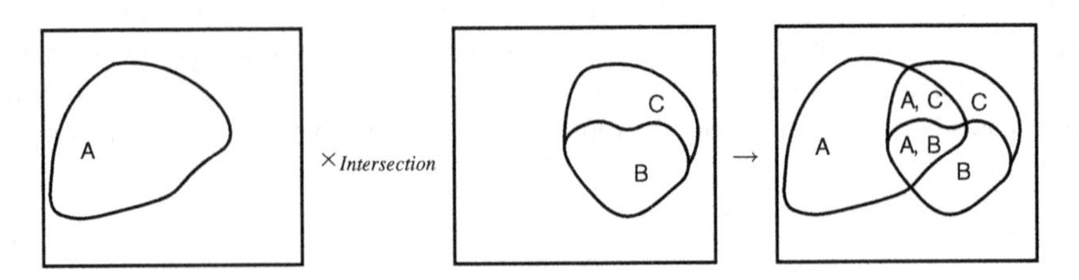

Fig. 9.6 The result of the *intersect* operation applied to two maps

Table 9.1 Assume Fig. 9.3a depicts two maps, each containing a single region. (a) is the label table for the map containing the region with identifier 1. (b) is the label table for the map containing the region with identifier 2. (c) is the label table for the result of the intersect operator applied to the first two maps. Figure 9.3b shows the resulting geometry with labels

(a)		(b)	
region_id: int	description: string	region_id: int	description: string
1	Region of interest	2	High temperature area

(c)		
region_id: int	description1: string	description2: string
1	Region of interest	
2		High temperature area
3	Region of interest	High temperature area

Algorithm 1: The algorithm to compute the label table for intersecting two maps.

Input: The label tables, M_L and P_L, for two maps M and P. A mapping A of labels from the original maps to labels in the result map. Lists L_M and L_P of labels from the original maps that exist in the result map. The map Q defined as the halfsegments resulting from the intersection operator applied to M and P.

Output: Label table Q_L

1 Create the label table Q_L;
2 **foreach** *region label in* L_M **do**
3   Compute a tuple for Q_L consisting of label attributes from L_M and default values for attributes from L_N;
4   Insert the new tuple in Q_L;
5 **end**
6 **foreach** *region label in* L_N **do**
7   Compute a tuple for Q_L consisting of label attributes from L_N and default values for attributes from L_M;
8   Insert the new tuple in Q_L;
9 **end**
10 **foreach** *entry in mapping A* **do**
11   The mapping entry consists of region labels $r_M, r_N \rightarrow r_Q$ for region labels in M, N, and Q, respectively;
12   Compute a tuple for Q_L consisting of label attributes from L_M and L_N with region_ids equal to r_M and r_N, respectively;
13   Insert the new tuple with region_id r_Q in Q_L;
14 **end**

mapping of labels from the input maps to labels in the result map. Therefore, the construction of the label table does not increase the time complexity of the operation since it simply iterates over each lists and mapping once.

9.4 The Relabel Operator

The relabel operation, as defined by the abstract model for maps, takes a map and a relabeling function which is then used to alter the labels of a map. A relabeling may have geometric consequences. For instance, relabeling a region to the exterior label effectively removes a region from a map. Relabeling a region such that its new label is identical to an adjacent region will merge the two regions, causing any boundaries that exist between them to disappear.

From an implementation perspective, defining a method for a user to create a relabeling function that can then be passed to an operation is non-trivial. Therefore, we approach the problem of relabeling from a different perspective. Our approach is to allow a user to modify a label table for a map, and then run a general relabeling operation that determines any geometric implications of the modifications to the label table, and updates the map geometry accordingly. This approach does not require a method of passing a relabeling function to the relabel operation, while not sacrificing any generality.

As was briefly mentioned in the examples above, the relabeling operation can have two geometric consequence: (i) labels may be removed from a label table, which implies that the region with that label has been deleted; and (ii) a label in the label table may be altered such that it is identical to another label in the label table, meaning that two regions have been merged into a single region (requiring any shared boundaries between the regions to be removed). Thus, our relabel operation must be able to detect and enforce these changes in the map geometry.

The relabel operation consists of two parts, a duplicate label identification part, and a geometric consistency enforcement part. The duplicate label identification part identifies labels in the label table that are identical, but that have different region_id values. This is achieved by computing a Cartesian product of a label table with itself, and then disregarding all tuples that are not identical except for their region_ids. The result is a relation containing pairs of identical labels that belong to different regions. This relation is traversed, and a mapping is constructed such that the region_id value of duplicate labels are mapped to the same region_id value. Furthermore, a list is constructed of all region_id values that exist in the label table.

Once the mapping of duplicate labels and the list of region_ids in the label table are constructed, they are used to determine changes to the map geometry. This is achieved by traversing the halfsegment sequence of the map geometry. For each halfsegment, the region label for the region above the halfsegment is checked against the label mapping and label list. If the region label is not in the label list, then the corresponding label has been removed from the label table, and the region should no longer exist in the map. Therefore, the region label for the halfsegment is changed to the exterior label. If the region label is in the label list, then the mapping is checked. If the label exists in the mapping, then it is changed to the label to which it is mapped. The same procedure is performed for the label of the region below the current halfsegment. Finally, once the above and below labels of the halfsegment have been processed, a final check is performed to ensure that the above and below labels are not identical.

If these labels are identical, then the halfsegment borders two regions with identical labels, and thus, is removed. Algorithm 2 summarizes the relabel algorithm.

Algorithm 2: The relabel algorithm.

```
   Input: A map M and its corresponding label table L_M.
   Output: Map M with the geometric changes implied by its label table L_M.
 1 Compute the Cartesian product L_M × L_M;
 2 Keep only the tuples in L_M × L_M that have different region_ids and identical labels;
 3 Create mapping A that maps region_ids for equivalent labels to the same region_id;
 4 Create list V of region_ids in L_M;
 5 for each halfsegment h in M do
 6     Get the above label a of h;
 7     Get the below label b of h;
 8     if a exists in mapping A then
 9         Replace a with the value to which it maps;
10     end
11     if a does not exists in list V then
12         Replace a with the exterior label;
13     end
14     if b exists in mapping A then
15         Replace b with the value to which it maps;
16     end
17     if b does not exists in list V then
18         Replace b with the exterior label;
19     end
20     if a = b then
21         Remove h from M;
22     end
23 end
```

Let M be an instance of the map2D type with r regions and n line segments. Computing a full cartesian product clearly has time complexity $O(r^2)$ since every region has a single label. This can be reduced $O(r \lg r)$ using indexing techniques or even sorting. A hash table is then used to construct the mapping from old labels to new labels. A hash table can be used to record the list of labels in the label table and provide existence checking for labels in constant time. Since each halfsegment must check its labels for existence, $O(n)$ hash table lookups will occur. Therefore, the algorithm has time complexity $O(r \lg r + n)$.

9.5 The Refine Operator

The refine operation, as was defined previously, relabels region faces such that every region in a map has only a single face. In other words, any regions that consist of multiple faces in a map will be altered such that each face is a single region in the map after a refine operation has been performed. In the implementation model of the Map Framework, this is achieved by altering the label table of a map such that two faces of a region will have unique labels, and every face of a region in a map will have a unique region_id. We compute this in two steps. First, the map geometry is modified so that each region face is identified by a unique region_id. Second, the label table is modified so that every region_id in the map geometry refers to a unique label in the label table.

The first step of the refine algorithm requires a technique known as *cycle walking*. In essence, given the left most halfsegment of a cycle (i.e., a region face in a map), it is possible to find each successive halfsegment that makes up the cycle of halfsegments in an efficient manner. First, we present the algorithm to walk the cycles in a *region*. We then extend the method to work on cycles in a map. We then discuss the second part of the refine operation

In the following discussion, we assume that halfsegments carry a boolean flag, denoted the *interior above flag*, that has a value of *True* if the interior of the region that a halfsegment bounds lies *above* (or to the left in the case of a vertical halfsegment) the halfsegment, and has a value of *False* otherwise. The interior above flag can be implemented with labels, but the use of a flag makes this discussion more concise.

9.5.1 Walking Cycles in Regions

Because this this section deals with regions, we will assume a halfsegment structure as defined for regions; i.e., a halfsegment contains a segment, a boolean flag indicating if it is a left or right halfsegment, and a boolean flag indicating if the interior of the region lies above or below the halfsegment. Furthermore, we develop this algorithm to determine the *component view* of a region. In essence, the algorithm determines which cycles in a region are outer cycles, which are hole cycles, and to which outer cycle each hole belongs. It also determines if a set of halfsegments satisfies the definition of a region. Although all of these properties are not required for the cycle walk portion of the refine algorithm, they are useful, so we include them. We assume that the input to our algorithm is a sequence of ordered halfsegments. If the input represents a region, the algorithm returns the region with its cyclic structure information; otherwise, the algorithm exits with an error message indicating the input sequence does not form a semantically correct region. We begin by providing a high level overview of the algorithm and then present the algorithm and provide a discussion of its details.

In general terms, the cycle walk algorithm must identify all cycles present in the halfsegment sequence, and classify each cycle as either an outer cycle or a hole cycle of a particular face [19]. To accomplish this, each halfsegment is *visited* once by the algorithm. Note that due to the definition of the type *region*, each segment belongs to exactly one cycle. When a halfsegment is visited, the algorithm marks the halfsegment indicating to which face and cycle it belongs, and whether that cycle is an outer cycle or a hole cycle. The algorithm does not alter the input when marking halfsegments, rather a parallel array to the input sequence is used to represent the cycle information. The algorithm visits halfsegments by stepping through the input list sequentially.

The first halfsegment in the input sequence will always be part of the outer cycle of a face, due to the definition of complex regions and the halfsegment ordering defined previously. Therefore, it can be visited and marked correctly. Once a halfsegment has been visited, it is possible to visit and correctly mark all other halfsegments in the cycle that it belongs to in a procedure which we denote as the *cycle walk*. Thus, all halfsegments that form the cycle to which the first halfsegment in the input sequence belongs are then visited. The algorithm then begins stepping through the remaining halfsegments. The next unvisited halfsegment encountered will be part of a new cycle. The algorithm then visits this new halfsegment. The algorithm can deduce whether this halfsegment is an outer cycle of a new face or a hole in an existing face by examining where the halfsegment lies in relation to already known cycles. To determine this, we use a plane sweep algorithm to step through the halfsegments. Thus, we can take advantage of the plane sweep status structure to find whether or not the current halfsegment lies in the interior of a previously visited face. Once the new halfsegment is visited, we perform a cycle walk from it. Then, the algorithm continues stepping through the input list until it reaches another unvisited halfsegment, visits it, and repeats this procedure. The algorithm is shown in Algorithm 3.

To properly describe the algorithm outlined in Algorithm 3, we introduce several notations. The function $info(h)$ for a given halfsegment h returns its cyclic information, that is, its *owning* cycle, and if part of a hole, its owning face. A cycle *owns* a halfsegment if the halfsegment is part of the boundary of the cycle, and a face owns a hole if the hole is inside the face. We define the function $NewCycle(h)$ to annotate h with a unique identifier for a new cycle. Let f be a halfsegment belonging to an outer cycle of a face. The function $Owns(h, f)$ annotates the halfsegment h to indicate that it belongs to a hole in the face that owns f. Finally, we employ the function $Visit(p)$ to mark a point p as having been visited. The function $Visited(p)$ is used to verify is point p was marked as visited already. Points are only marked as visited when a halfsegment with dominating point p has been visited during the cycle walk. We mark points as being visited in order to identify the special case of a hole cycle that meets the outer cycle of a face at a point. The function $Visited(h)$ is used to verify if halfsegment h has been visited already. A halfsegment has been visited if it has been annotated with face/hole information. For a halfsegment h, we can directly compute its corresponding right (left) halfsegment h_b, which we call its *brother* by switching its boolean flag indicating which end point is dominant. We define the next halfsegment in the cycle to which h belongs as h_+ such that the dominating endpoint of h_b is equal to the dominating point $dp(h_+)$ and $h_+ \neq h_b$ and h_+ is the first halfsegment encountered when rotating h_b clockwise (in an outer cycle) or counter-clockwise (in a hole cycle) around its dominating point. The previous halfsegment in the cycle is similarly defined as h_-.

Algorithm 3: The algorithm for deriving the component view of a region.

```
   Input: Sequence of unannotated halfsegments H
   Output: Sequence H with fully annotated halfsegments
 1 while not end of sweep do
 2     Advance sweep line to h. h is the left-most halfsegment yet to be annotated;
 3     Using sweep line status, determine h as part of an outer cycle or a hole cycle;
 4     NewCycle(h);
 5     Visit(dp(h));
 6     if h belongs to a hole then
 7         Using sweep line status, retrieve halfsegment f from its owning outer cycle;
 8         Owns(h, f);
 9         Set cycle walk mode to use counter-clockwise adjacency;
10     else
11         Set cycle walk mode to use clockwise adjacency;
12     end
       /* Begin walking the cycle */
13     c ← h+;
14     while c ≠ h do
15         if Visited(dp(c)) then
16             q ← c;
17             c ← c−;
18             NewCycle(c);
19             Owns(c, h);
20             while dp(c) ≠ dp(q) do
                   /* Trace back anchored hole */
21                 info(c−) ← info(c);
22                 c ← c−;
23             end
24         else
25             info(c) ← info(h);
26             Visit(dp(c));
27             c ← c+;
28         end
29     end
30 end
```

9.5.2 *Classifying Outer and Hole Cycles*

By using a sweep line, the algorithm steps through the halfsegment sequence to find the smallest unannotated halfsegment h, create a new cycle for this halfsegment, and mark its dominating point as visited (lines 2–5). At this point, the algorithm needs to determine whether h belongs to a hole cycle (line 6) or an outer cycle (line 10). If a cycle is identified as a hole cycle, the outer cycle to which it belongs must also be identified (line 7–9), and the cycle must be walked using counter-clockwise adjacency of halfsegments (line 9).

Recall that the plane sweep algorithm maintains the sweep line status structure, which is a ordered list of *active* segments, such that it provides a consistent view of all halfsegments that currently intersect the sweep line, up to the current *event* (the addition or removal of a halfsegment). By examining the halfsegment directly below a halfsegment h in the sweep line status, we can determine whether h is a part of an outer cycle or a hole cycle of an existing face. In other words, if halfsegment g is directly below halfsegment h in the sweep line status structure and the interior-above flag of g is set to *true*, it follows that h is either in the interior of the cycle to which g belongs, or h is part of the cycle to which g belongs. Recall that as soon as a halfsegment is classified as being a part of a hole or face, the cycle to which it belongs is walked and all other halfsegments in that cycle are marked accordingly (lines 14–29). Therefore, if a halfsegment belongs to the same cycle as any halfsegment that has been previously encountered by the sweep line, it is already known to which face and/or hole cycle it belongs. Furthermore, all halfsegments that are less than a given halfsegment in halfsegment order have already been classified. Therefore, we can determine if an unmarked halfsegment belongs to a hole or outer cycle by examining the halfsegment immediately below it in the sweep line status structure.

From the definition of a face, the outer cycle of a face of a region always covers (encloses) all of its hole cycles. This means that the smallest halfsegment of this face is always a part of the outer cycle. This is also true for the entire region object where the smallest halfsegment in the ordered sequence is always a part of the first outer cycle of the first face. Furthermore, due to the order relation of halfsegments and the cyclic structure of a polygon, the smallest halfsegment of a face will always be a left halfsegment with the interior of the face situated above it. Thus, when we process this halfsegment, we set its interior-above flag to indicate this fact. Since we have classified this cycle as an outer cycle, we can walk the cycle and set the interior-above flag for all halfsegments of this cycle. For example, Fig. 9.7a illustrates the case where the smallest halfsegment of the sequence is processed and the cycle is classified as an outer cycle.

Once the first outer cycle of a face in a region has been processed, we continue to process halfsegments that have not yet been classified based on the plane sweep status structure. Figure 9.7b shows an example. Here, we add/remove visited halfsegments into/from the sweep line status in sequence ordered up to the smallest unvisited halfsegment k. This halfsegment *must* be the start of a new cycle that we must now classify. We know k is the start of a new cycle because all halfsegments of an existing cycle that include a halfsegment j such that $j < k$ must have been marked as visited by the walking process. Once we reach this new cycle represented by its starting

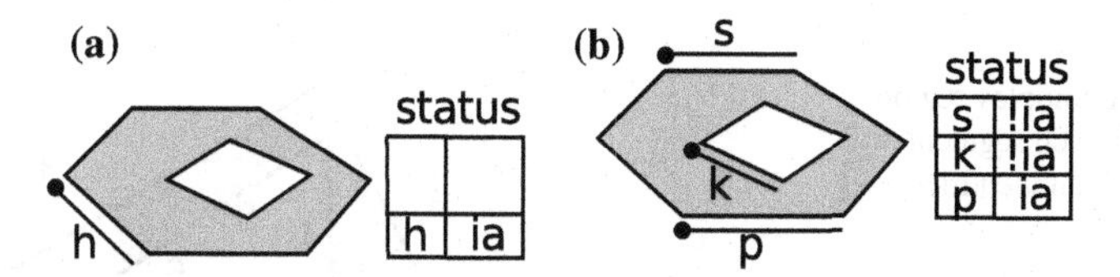

Fig. 9.7 Processing halfsegments in a region. **a** Processing the smallest halfsegment h of the sequence. **b** Processing halfsegment k of a cycle. *ia* indicates the value of the interior above flag

halfsegment k, we add this halfsegment into the sweep line status. We classify the type of cycle k belongs to by examining the interior-above flag of the halfsegment p (its predecessor) which was already visited and sits immediately below k in the sweep line status structure. If the predecessor indicates that the interior of the face is above it (the interior-above attribute of p is set to true), then k lies in the interior of the cycle to which p belongs; thus, k must be part of a hole cycle and the interior of the face to which k belongs must lie below k. If the interior-above flag of p indicates that the interior of the face to which p belongs is below p, then the current halfsegment k must be part of an outer cycle of a new face. In the case that there is no predecessor, then the current halfsegment must be a part of an outer cycle of a new face, because it does not lie in the interior of any other face's outer cycle. Once the cycle is classified as either an outer cycle of a new face or a hole cycle of an existing face, the cycle walking procedure is carried out to determine all halfsegments that belong to the cycle.

9.5.3 *Walking Cycles*

In general terms, we use the phrase *walking a cycle* to indicate the traversal of a cycle such that each halfsegment that forms the cycle is visited. Furthermore, the halfsegments in such a traversal are visited in the order in which they appear in the cycle. In other words, given a halfsegment h, all halfsegments in the cycle to which h belongs are found by repeatedly finding h_+ until the the original halfsegment is encountered again. For example, when walking the outer cycle of the region in Fig. 9.8 in clockwise order beginning from S_1, the halfsegments would be encountered in the order 1_l, 1_r, 3_r, 3_l, 2_r, 2_l. The two main challenges to this portion of the algorithm are (i) to identify cycles correctly such that they correspond to the unique representation of a region as stated in the definition of complex regions, and (ii) to achieve this efficiently.

When a halfsegment h is encountered by the algorithm that has not yet been classified, it is classified as belonging to a hole or outer cycle in line 6. If h belongs to an outer cycle, then the cycle walk portion of the algorithm in lines 14–29 is executed. Due to the halfsegment ordering and the definition of regions, the smallest unvisited halfsegment in the input sequence that the plane sweep encounters is always a left halfsegment of an outer cycle of a face and the interior of that face always lies

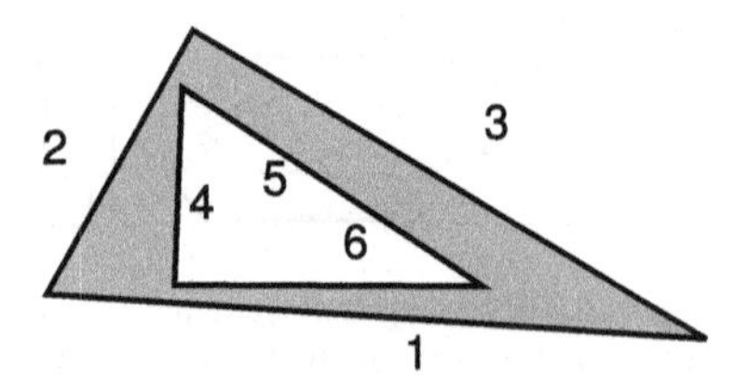

Fig. 9.8 A complex region object with each segment labeled with an identifier

above the halfsegment. If we rotate h_b clockwise around its dominating point, it will intersect the interior of the face. Thus, the first halfsegment encountered when rotating h_b clockwise around its dominating point will be part of the outer cycle of the same face (except for a special case discussed below) and will be h_+. We know this to be true because if we find h_+ in this fashion and it turns out to be part of another face, then two faces would intersect, which is prohibited by the definition of complex regions. It follows that each successive halfsegment in the outer cycle can be found by rotating the brother of the current halfsegment clockwise around its dominating point, because the location of the interior relative to the halfsegment can always be deduced based on the previous halfsegment encountered in the cycle walk.

One special case occurs when walking outer cycles: the existence of a hole in a face that meets the outer cycle at a point. When walking an outer cycle that contains such a hole, the halfsegments that form the hole will be classified as being part of the outer cycle using the procedure just described. In order to remedy this, we mark each point that is a dominating point of a halfsegment encountered during the cycle walk (line 26). Each time we find a new halfsegment that is part of an outer cycle, we first check if its dominating point has been visited yet (line 15). If it has been visited, then we know that we have encountered that point before, and a hole cycle that meets the outer cycle must have been discovered. When this happens, we loop backwards over the cycle until we find the halfsegment whose dominating point has been visited twice (lines 15–23). The halfsegments forming the hole are then marked as such. We denote such a hole an *anchored hole*, since it is anchored to an outer cycle at a point. The remainder of the outer cycle is then walked.

Walking a hole is identical to walking an outer cycle, except that a counter-clockwise rotation from h_b is used to find h_+. A counter-clockwise rotation is required because the interior of the face is intersected by h_b when rotating h_b counter-clockwise around its dominating point. When walking holes, the special case exists that two holes may meet at a point. Thus, we employ the same strategy to detect this case as we did with the special case of a hole meeting a face (lines 15–18).

9.5.4 Walking Cycles in Maps

Walking cycles in a map is similar to walking cycles in a region. However, there are some differences. First, because regions in a map can share halfsegments that separate them, we cannot simply mark a halfsegment as being visited. Instead, we must identify if we have visited the halfsegment when walking the cycles of the region above the halfsegment, or below the halfsegment. Therefore, two parallel arrays are required to store this information, one to mark the above side of halfsegments that have been visited, and the other to mark the below side. The second difference is that the purpose of the refine algorithm is to give every face in a map a unique label. Therefore, we must remember the region_id of each face we have visited, and change the region_id of any face that has the same region_id of a face that has already been visited to a new, unique region_id. We achieve this by maintaining a mapping of region_id values from the original map to region_id values in the refined map. At the first halfsegment encountered for each cycle, we check the mapping to see if the region_id for the new cycle is in the mapping. If not, then it is the first time that region_id has been encountered. In this case, we add the region_id to the mapping such that it maps to itself. If the region_id is already in the mapping, then we add the old region_id to the mapping such that it maps a new region_id x; furthermore, as the cycle corresponding to the face is being walked, the region_id for every halfsegment in the cycle is changed to x. These are the only modifications required for the cycle walking algorithm presented above to be applied to the refine operation of maps.

9.5.5 Extending the Label Table

The second part of the refine operation alters the label table for a map so that it is consistent with the map geometry created in the cycle walking portion of the refine algorithm. In order to ensure that region labels remain unique, the label table for the map being refined is altered so that it contains an additional column, which we will call *ref*, of type integer. This column is given a default value, such as -1, that is not used as a region_id value. Once the label table has this new column, then the tuples corresponding to the labels of the new regions (which previously were faces of regions) in the map must be inserted. This is achieved using the label mapping computed during the cycle walk. Recall that the cycle walk portion of the algorithm returns a mapping A from region_ids belonging to the map before the cycle walk portion of the algorithm has taken place, to region_ids belonging to the map after the cycle walk portion of the algorithm has taken place. Therefore, for each region_id i in A that maps to a new region_id j that is not in the label table, a tuple with region_id j must be inserted into the label table with identical attribute values as the tuple with region_id i, except that the *ref* column must contain a different value than the tuple with region_id i. Therefore, the value of j is used in the *ref* column, since it is unique. Once every entry in the mapping has been processed, the refine operation is complete. Algorithms 4 and 5 summarize this.

From the discussion of the cycle walk portion of the algorithm, it is clear that the plane sweep and altering of halfsegment has time complexity $O(n \lg n)$ for a map with n halfsegments. The addition of a mapping of region labels to new region labels can be implemented using hash table, and thus has a complexity of $O(r)$ for a map containing r faces. Finally, the portion of the algorithm that alters the label table must alter each existing tuple, plus insert a new tuple for each new region identified in the cycle walk portion of the algorithm. Thus for a map containing r faces, this takes $O(r)$ time since each tuple is modified or inserted once. Therefore, the time complexity for the refine algorithm is $O(n \lg n + r + r)$. Since the number of halfsegments in a map is typically much greater than the number of regions, this tends to be dominated by first term.

Algorithm 4: The cycle walk portion of the refine algorithm.

```
   Input: A map M.
   Output: M modified according to the refine operation, and a label mapping A.
 1 Initialize sweep line structures;
 2 while not end of sweep do
 3     Get next halfsegment h with an unvisited side;
 4     if above side of h has not yet been visited then
 5         Check mapping A for above label a of h;
 6         if a is in mapping A then
 7             Set a to map to new region_id c in A;
 8             Change a to c in h;
 9             Walk the cycle as in Algorithm 3, marking the appropriate side of halfsegments as
                 being visited and changing appropriate labels;
10         else
11             Set a to map to a in A;
12             Walk the cycle as in Algorithm 3, marking the appropriate side of halfsegments as
                 being visited;
13         end
14     end
15     if below side of h has not yet been visited then
16         Check mapping A for below label b of h;
17         if b is in mapping A then
18             Set b to map to new region_id c in A;
19             Change b to c in h;
20             Walk the cycle as in Algorithm 3, marking the appropriate side of halfsegments as
                 being visited and changing appropriate labels. Use counter-clockwise ordering
                 around points.;
21         else
22             Set b to map to b in A;
23             Walk the cycle as in Algorithm 3, marking the appropriate side of halfsegments as
                 being visited. Use counter-clockwise ordering around points.;
24         end
25     end
26 end
```

Algorithm 5: The label table modification portion of the refine algorithm.

```
   Input: A label table L_M for map M and a mapping A from the cycle walk portion of the
          refine algorithm.
   Output: Label table L_M modified according to the label mapping A.
 1 Add a new column named ref to L_M with a default value;
 2 for each entry a → b in A do
 3     if a! = b then
 4         Construct tuple t identical to the tuple with region_id a in L_M;
 5         Set the region_id of t to b;
 6         Set the ref column value of t to b;
 7     end
 8 end
```

9.6 Combining Operations to Form New Operations

It is possible to combine the previous three operations to create new operations. For instance, a common operation for maps is to overlay two maps such that the result map only contains the intersecting portions of regions in the original maps. We denote this operation $overlay_2$ (to differentiate it from the term *overlay* which typically refers to the intersect operation described above), and show an example in Fig. 9.9.

In order to compute the $overlay_2$, we simply combine the *intersect* and *relabel* operations defined previously. Given two argument maps, M and P, we first compute the intersect operation, resulting in map Q. Recall that the label table for the map Q will have all columns from the label tables for M and P; furthermore, the column names corresponding to columns that came from the first argument map, M, will have a 1 appended, and the column names corresponding to columns that came from the second argument map P, will have 2 appended. In order to compute the $overlay_2$, we must remove all regions from Q that cover an area that is covered by a region in only a single argument map. We can identify those regions due to the fact the the attributes in the label table for all attributes from a single argument region will have a default value. Therefore, we must simply remove the tuples that satisfy this property. Once those tuples are removed, the relabel operation is executed over Q, which enforces the geometric consequences of removing tuples from the label table of Q (i.e., the corresponding regions are removed). After the relabel, Q contains the result of the $overlay_2$ operation applied to M and P.

9.7 Querying Maps

Using the data model defined in this chapter, we show how to construct queries over a map2D type in a relational DBMS. We will use the running example of a table that contains a map2D type, and associated attribute tables for the maps. We will use the example depicted in Fig. 9.1, which is repeated as Fig. 9.10 for convenience.

Traditional spatial types have two types of queries associated with them, spatial queries and attribute queries [8, 14, 27, 30, 33]. Maps allow for new types of queries

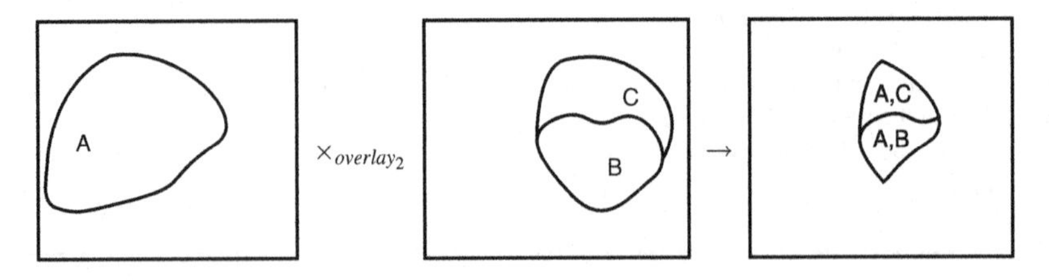

Fig. 9.9 The result of the $overlay_2$ operation applied to two maps

due their more complex structure and the integration of attribute data in the data type definition. Thus, instead of being defined merely as a geometry, a map has four components which may be involved in queries: (i) a map geometry, (ii) components in the form of *regions* that compose the geometry, (iii) attributes associated with each region, and (iv) attributes associated with the entire map. Therefore, we classify map queries into four classes of queries: *map queries*, which look at the map as a whole; *map attribute queries*, which are similar to attribute queries over traditional spatial objects in that they deal with attributes stored separately from a map object; *component attribute queries*, which deal with the attributes associated with regions in a map; and *component queries*, which deal with the regions that compose a map.

We begin by considering the creation of maps, then we discuss the types of map queries and show how they may be implemented in SQL. Some extensions are required for SQL, and those are introduced as needed.

9.7.1 Creating Maps

In this section we show how we can extend the SQL *CREATE* and *INSERT* constructs to accommodate the creation of tables containing map2D columns and tuples containing map2D instances. We assume that the database is aware of the implementation data model of map2D as discussed previously, specifically the type *map2D*. We now show how a user can create the tables depicted in Fig. 9.10.

Consider a user who wishes to create a map in a database. We assume that this user begins with an empty database (i.e., a database containing no tables) and performs the necessary steps to create a map2D instance within the database. The first step the user must take in order to create a map is to create a table in which the map will be

MapTable

ID: int	Name: str	Geom: map2D
1	Map_A	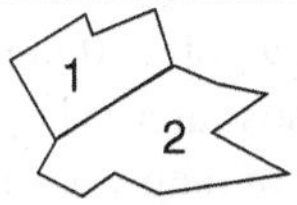 1 2
2	Map_B	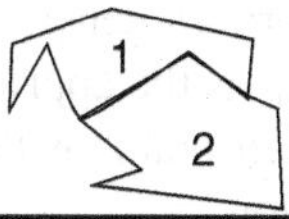 1 2

Map1AttributeTable

region_id: int	Crop: str
1	Wheat
2	Corn

Map2AttributeTable

region_id: int	Avg Temp: int	Avg Rain: int
1	98	14
2	86	22

Fig. 9.10 A relation containing a map2D column and the associated label tables. The table Map1AttributeTable is associated with the map with ID equal to 1 in the table MapTable, and the table Map2AttributeTable is associated with the map with ID equal to 2 in the table MapTable

stored. Since the database is aware of the map2D type, we can simply use a create table command as shown below:

Statement 1 *CREATE TABLE MapTable(ID: int, Name: str, Geom: map2D)*

The above create statement creates an empty table. In order to insert a tuple into the table, we require a map2D instance, which can be an empty map, and an associated label table. A map must have an associated label table, thus, we assume that a map2D constructor will take a label table name as an argument, and associate the map2D geometry with the label table and ensure that the map2D type constraints are satisfied (for instance, recall that we impose the restriction on a label table that it must contain a column of type integer with the name *region_id* that will be used to associate each entry in the label table with a region in the map). For example, to create the table named *Map1AttributeTable* in Fig. 9.10, a user would use the following statement:

Statement 2 *CREATE TABLE Map1AttributeTable(region_id: int, crop: str)*

A user creates a map whose regions are associated with the thematic values in a label table through an *INSERT* statement. We assume the existence of a map constructor, which we indicate as *map2D(string labelTable)*, that takes the name of a label table as an argument and returns an empty map object that is associated with the label table. In other words, the map object stores the name of the label table internally.

Statement 3

INSERT INTO MapTable
VALUES(1, 'Map_A', map2D('Map1AttributeTable'))

9.7.2 General Map Queries

Map queries, are queries that consider a map as a whole, often involving map operations. For example, queries that return all maps whose area is larger than one thousand square feet, or that calculate the overlay of a pair of maps, or that return maps based on a topological relationship fall into this category. In general, queries of this type are expressed in a straightforward manner as long as the appropriate map operation is defined. For example, in order to find all maps in a table with an area greater than one thousand, the user would use an *area* operation:

Query 5

SELECT Geom
FROM MapTable
WHERE Geom.area() > 1000

Queries involving map operations typically express the operation in the *SELECT* clause, and the use the other clauses to identify the maps involved in the operation:

Query 6

SELECT Overlay(M1.Geom, M2.Geom)
FROM MapTable M1, MapTable M2
WHERE M1.ID = 1 AND M2.ID = 2

Similarly, topological predicates between maps can be expressed. For example, to find all maps that satisfy the topological predicate *meet* (the maps share a boundary portion but their interiors do not intersect), a user would issue the following:

Query 7

SELECT M1.Geom
FROM MapTable M1, MapTable M2
WHERE meet(M1.Geom, M2.Geom)

Note that these queries can be expressed in a relatively straightforward manner, since they are similar in concept to traditional spatial queries over the traditional spatial types. In other words, we are querying the spatial object as a whole, and simply using operations and predicates that take instances of the spatial types as arguments.

9.7.3 Map Attribute Queries

Map attribute queries refer to queries that operate on the attributes associated with a map as a whole. In other words, queries that operate on the attributes of the *MapTable* in Fig. 9.10 other than the *Geom* attribute are map attribute queries. These queries are traditional SQL queries. For example, a user may wish to find the *ID* of the map named "Map_A":

Query 8

SELECT ID
FROM MapTable
WHERE Name = 'Map_A'

9.7.4 Component Attribute Queries

Component attribute queries are unique to the map2D type since they are querying attributes related to geometric components of an instance of a map2D type in a table column. The ability to associate attribute data with the components of spatial objects is unique to spatial partitions in the DBMS setting. For example, a user may wish to know the average temperature of regions whose average rainfall is greater than 20 inches per year for regions in *Map_B* in Fig. 9.10. Such a query could be expressed as:

Query 9

SELECT Avg_Temp
FROM Map2AttributeTable
WHERE Avg_Rain > 20

This works if the user is aware of the name of a label table for a given map. If the user does not know this information, we must provide a means of accessing a map's label table through the map type itself. This requires an extension to SQL. Note that other implementations of a map2D type in a DBMS might not require an SQL extension, but the concept embodied by this extension will still be required. We introduce an operation called *GetLabelTable* which takes a map2D instance as an argument and returns the relation consisting of the label table associated with the given map. Given such an operation, we can then express component attribute queries given instances of the map2D type:

Query 10

SELECT Avg_Temp
FROM GetLabelTable(SELECT GEOM FROM MapTable WHEREID = 2)
WHERE Avg_Rain > 20

Clearly, this query is simplistic in that we specified the *ID* of the map whose table we wanted to query. If we do not know which particular map we want to query, we must use a *correlated subquery* approach in which the label table for each map is examined independently. For example, assume that a user wishes to find the maps in the database that contain a region with average rainfall that is greater than 20 in. In this case, we do not specify which map the regions may exist in, rather, we examine the regions of all maps. We construct the query as:

Query 11

```
SELECT Geom
FROM MapTable M1
WHERE EXISTS
    ( SELECT*
     FROM GetLabelTable( M1.Geom, integer : Avg_Rain )
     WHERE Avg_Rain > 20 )
```

One issue that arises is that each map may have label tables of different structure; thus, not all label tables will contain a column of type *int* with name *Avg_Rain*. Therefore, we specify in the *GetLabelTable* the name and type of columns that must exist in order for the subquery to be executed. If the column does not exist in the label table, the subquery returns an empty relation.

A second example returns all maps containing regions in a column named *crop* that have a value of *wheat*:

Query 12

```
SELECT Geom
FROM MapTable M1
WHERE EXISTS
    ( SELECT*
     FROM GetLabelTable( M1.Geom, string : Crop )
     WHERE crop = 'wheat' )
```

9.7.5 Component Queries

Component queries are similar in concept to component attribute queries, but deal with the region geometries contained in a map instead of the region attributes. Because the map geometry is stored in the column type *map2D*, we cannot use a correlated subquery concept similar to the approach used in component attribute queries. Instead, we must provide access to the regions in a map in a manner which corresponds to SQL and relational constructs. We express this class of queries in SQL by defining the *Expand* function. We define the *Expand* function to take a map2D instance and return a relation consisting of singleton maps such that each singleton map represents a region in the original map. Furthermore, each singleton map will have a label table associated with it containing a single region label. For example, if a user wishes to return all regions from a map that have an area greater than one thousand square miles they could use the *Expand* operator in conjunction with an *area* operator defined for regions as follows:

Query 13

SELECT M1.Geom
FROM Expand(SELECT Geom from MapTable) as M1
WHERE area(M1.Geom) > 1000

We can also use the Expand operator in correlated subqueries to identify maps that contain regions that satisfy some predicate. For example, to return all maps that contain a region with an area greater than one thousand square miles, a user could write:

Query 14

SELECT M1.Geom
FROM MapTable M1
WHERE EXISTS
(SELECT M2.Geom
FROM Expand(M1.Geom) as M2
WHERE area(M2.Geom) > 1000)

9.8 Summary

In this chapter, we describe an implementation model for spatial partitions that includes data type implementation details and algorithms for implementing operations over spatial partitions. Furthermore, we showed how such a model can be integrated into a DBMS, and described the types of queries that can then be issued over spatial partitions and provided examples of those types. Some of the geometric algorithms in the chapter are implemented in the freely available Pyspatiotemporalgeom Library [22, 23].

References

1. Arge, L., Procopiuc, O., Ramaswamy, S., Suel, T., Vitter, J.S.: Scalable sweeping-based spatial join. In: VLDB, vol. 98, pp. 570–581. Citeseer (1998)
2. Audet, S., Albertsson, C., Murase, M., Asahara, A.: Robust and efficient polygon overlay on parallel stream processors. In: Proceedings of the 21st ACM SIGSPATIAL International Conference on Advances in Geographic Information Systems, pp. 304–313. ACM (2013)
3. Balaban, I.J.: An optimal algorithm for finding segments intersections. In: Proceedings of the eleventh annual symposium on Computational geometry, pp. 211–219. ACM (1995)
4. Bentley, J.L., Ottmann, T.A.: Algorithms for reporting and counting geometric intersections. Computers, IEEE Transactions on **100**(9), 643–647 (1979)

5. Chazelle, B., Edelsbrunner, H.: An optimal algorithm for intersecting line segments in the plane. Journal of the ACM (JACM) **39**(1), 1–54 (1992)
6. Chazelle, B., Edelsbrunner, H., Guibas, L.J., Sharir, M.: Algorithms for bichromatic line-segment problems and polyhedral terrains. Algorithmica **11**(2), 116–132 (1994)
7. De La Briandais, R.: File searching using variable length keys. In: Papers presented at the the March 3-5, 1959, western joint computer conference, pp. 295–298. ACM (1959)
8. Egenhofer, M.J.: Spatial sql: A query and presentation language. Knowledge and Data Engineering, IEEE Transactions on **6**(1), 86–95 (1994)
9. Erwig, M., Schneider, M.: Partition and conquer. In: Spatial Information Theory A Theoretical Basis for GIS, pp. 389–407. Springer (1997)
10. Erwig, M., Schneider, M.: Formalization of advanced map operations. In: 9th Int. Symp. on Spatial Data Handling, vol. 8, pp. 3–17 (2000)
11. Fredkin, E.: Trie memory. Communications of the ACM **3**(9), 490–499 (1960)
12. Goodrich, M.T.: Intersecting line segments in parallel with an output-sensitive number of processors. SIAM Journal on Computing **20**(4), 737–755 (1991)
13. Goodrich, M.T., Ghouse, M.R., Bright, J.: Sweep methods for parallel computational geometry. Algorithmica **15**(2), 126–153 (1996)
14. Güting, R.H.: An introduction to spatial database systems. The VLDB Journal The International Journal on Very Large Data Bases **3**(4), 357–399 (1994)
15. Güting, R.H., Schilling, W.: A practical divide-and-conquer algorithm for the rectangle intersection problem. Information Sciences **42**(2), 95–112 (1987)
16. Kriegel, H.P., Brinkhoff, T., Schneider, R.: The combination of spatial access methods and computational geometry in geographic database systems. In: Advances in Spatial Databases, pp. 5–21. Springer (1991)
17. Mairson, H.G., Stolfi, J.: Reporting and counting intersections between two sets of line segments. In: Theoretical foundations of computer graphics and CAD, pp. 307–325. Springer (1988)
18. McKenney, M.: Map algebra: A data model and implementation of spatial partitions for use in spatial databases and geographic information systems. Ph.D. thesis, University of Florida (2008)
19. McKenney, M.: Region extraction and verification for spatial and spatio-temporal databases. In: Scientific and Statistical Database Management, pp. 598–607. Springer (2009)
20. McKenney, M.: Geometric and thematic integration of spatial data into maps. In: Information Reuse and Integration (IRI), 2010 IEEE International Conference on, pp. 201–206. IEEE (2010)
21. McKenney, M.: Algorithms for spatial data integration. Recent Trends in Information Reuse and Integration pp. 257–272 (2012)
22. McKenney, M.: Pyspatiotemporalgeom package. https://pypi.python.org/pypi/pyspatiotemporalgeom/ (2016). Version 0.2, Accessed: 2016-06-22
23. McKenney, M.: Pyspatiotemporalgeom source code. https://bitbucket.org/marmcke/pyspatiotemporalgeom/ (2016). Accessed: 2016-06-22
24. McKenney, M., De Luna, G., Hill, S., Lowell, L.: Geospatial overlay computation on the gpu. In: Proceedings of the 19th ACM SIGSPATIAL International Conference on Advances in Geographic Information Systems, pp. 473–476. ACM (2011)
25. McKenney, M., Schneider, M.: Advanced operations for maps in spatial databases. Progress in Spatial Data Handling pp. 495–510 (2006)
26. Nievergelt, J., Preparata, F.P.: Plane-sweep algorithms for intersecting geometric figures. Communications of the ACM **25**(10), 739–747 (1982)
27. Opengis implementation specification for geographic information - simple feature access - part 2: Sql option (2010)
28. Ottmann, T., Wood, D.: Space-economical plane-sweep algorithms. Computer vision, graphics, and image processing **34**(1), 35–51 (1986)
29. Palazzi, L., Snoeyink, J.: Counting and reporting red/blue segment intersections. CVGIP: Graphical Models and Image Processing **56**(4), 304–310 (1994)

30. Rigaux, P., Scholl, M., Voisard, A.: Spatial databases: with application to GIS. Morgan Kaufmann (2001)
31. Shamos, M.I., Hoey, D.: Geometric intersection problems. In: Foundations of Computer Science, 1976., 17th Annual Symposium on, pp. 208–215. IEEE (1976)
32. Shcneider, M.: Spatial Data Types for Database Systems. Springer–Verlag (1995)
33. Shekhar, S., Chawla, S.: Spatial databases: a tour, vol. 2003. prentice hall Upper Saddle River, NJ (2003)
34. Vaishnavi, V.K., Wood, D.: Rectilinear line segment intersection, layered segment trees, and dynamization. Journal of Algorithms **3**(2), 160–176 (1982)

Printed in the United States
By Bookmasters